ENGINEERING DOCUMENTATION CONTROL PRACTICES AND PROCEDURES

MECHANICAL ENGINEERING

A Series of Textbooks and Reference Books

Editor

L. L. Faulkner

Columbus Division, Battelle Memorial Institute
and Department of Mechanical Engineering
The Ohio State University
Columbus, Ohio

1. *Spring Designer's Handbook*, Harold Carlson
2. *Computer-Aided Graphics and Design*, Daniel L. Ryan
3. *Lubrication Fundamentals*, J. George Wills
4. *Solar Engineering for Domestic Buildings*, William A. Himmelman
5. *Applied Engineering Mechanics: Statics and Dynamics*, G. Boothroyd and C. Poli
6. *Centrifugal Pump Clinic*, Igor J. Karassik
7. *Computer-Aided Kinetics for Machine Design*, Daniel L. Ryan
8. *Plastics Products Design Handbook, Part A: Materials and Components; Part B: Processes and Design for Processes*, edited by Edward Miller
9. *Turbomachinery: Basic Theory and Applications*, Earl Logan, Jr.
10. *Vibrations of Shells and Plates*, Werner Soedel
11. *Flat and Corrugated Diaphragm Design Handbook*, Mario Di Giovanni
12. *Practical Stress Analysis in Engineering Design*, Alexander Blake
13. *An Introduction to the Design and Behavior of Bolted Joints*, John H. Bickford
14. *Optimal Engineering Design: Principles and Applications*, James N. Siddall
15. *Spring Manufacturing Handbook*, Harold Carlson
16. *Industrial Noise Control: Fundamentals and Applications*, edited by Lewis H. Bell
17. *Gears and Their Vibration: A Basic Approach to Understanding Gear Noise*, J. Derek Smith
18. *Chains for Power Transmission and Material Handling: Design and Applications Handbook*, American Chain Association
19. *Corrosion and Corrosion Protection Handbook*, edited by Philip A. Schweitzer
20. *Gear Drive Systems: Design and Application*, Peter Lynwander

21. *Controlling In-Plant Airborne Contaminants: Systems Design and Calculations,* John D. Constance
22. *CAD/CAM Systems Planning and Implementation,* Charles S. Knox
23. *Probabilistic Engineering Design: Principles and Applications,* James N. Siddall
24. *Traction Drives: Selection and Application,* Frederick W. Heilich III and Eugene E. Shube
25. *Finite Element Methods: An Introduction,* Ronald L. Huston and Chris E. Passerello
26. *Mechanical Fastening of Plastics: An Engineering Handbook,* Brayton Lincoln, Kenneth J. Gomes, and James F. Braden
27. *Lubrication in Practice: Second Edition,* edited by W. S. Robertson
28. *Principles of Automated Drafting,* Daniel L. Ryan
29. *Practical Seal Design,* edited by Leonard J. Martini
30. *Engineering Documentation for CAD/CAM Applications,* Charles S. Knox
31. *Design Dimensioning with Computer Graphics Applications,* Jerome C. Lange
32. *Mechanism Analysis: Simplified Graphical and Analytical Techniques,* Lyndon O. Barton
33. *CAD/CAM Systems: Justification, Implementation, Productivity Measurement,* Edward J. Preston, George W. Crawford, and Mark E. Coticchia
34. *Steam Plant Calculations Manual,* V. Ganapathy
35. *Design Assurance for Engineers and Managers,* John A. Burgess
36. *Heat Transfer Fluids and Systems for Process and Energy Applications,* Jasbir Singh
37. *Potential Flows: Computer Graphic Solutions,* Robert H. Kirchhoff
38. *Computer-Aided Graphics and Design: Second Edition,* Daniel L. Ryan
39. *Electronically Controlled Proportional Valves: Selection and Application,* Michael J. Tonyan, edited by Tobi Goldoftas
40. *Pressure Gauge Handbook,* AMETEK, U.S. Gauge Division, edited by Philip W. Harland
41. *Fabric Filtration for Combustion Sources: Fundamentals and Basic Technology,* R. P. Donovan
42. *Design of Mechanical Joints,* Alexander Blake
43. *CAD/CAM Dictionary,* Edward J. Preston, George W. Crawford, and Mark E. Coticchia
44. *Machinery Adhesives for Locking, Retaining, and Sealing,* Girard S. Haviland
45. *Couplings and Joints: Design, Selection, and Application,* Jon R. Mancuso
46. *Shaft Alignment Handbook,* John Piotrowski

47. *BASIC Programs for Steam Plant Engineers: Boilers, Combustion, Fluid Flow, and Heat Transfer,* V. Ganapathy
48. *Solving Mechanical Design Problems with Computer Graphics,* Jerome C. Lange
49. *Plastics Gearing: Selection and Application,* Clifford E. Adams
50. *Clutches and Brakes: Design and Selection,* William C. Orthwein
51. *Transducers in Mechanical and Electronic Design,* Harry L. Trietley
52. *Metallurgical Applications of Shock-Wave and High-Strain-Rate Phenomena,* edited by Lawrence E. Murr, Karl P. Staudhammer, and Marc A. Meyers
53. *Magnesium Products Design,* Robert S. Busk
54. *How to Integrate CAD/CAM Systems: Management and Technology,* William D. Engelke
55. *Cam Design and Manufacture: Second Edition; with cam design software for the IBM PC and compatibles, disk included,* Preben W. Jensen
56. *Solid-State AC Motor Controls: Selection and Application,* Sylvester Campbell
57. *Fundamentals of Robotics,* David D. Ardayfio
58. *Belt Selection and Application for Engineers,* edited by Wallace D. Erickson
59. *Developing Three-Dimensional CAD Software with the IBM PC,* C. Stan Wei
60. *Organizing Data for CIM Applications,* Charles S. Knox, with contributions by Thomas C. Boos, Ross S. Culverhouse, and Paul F. Muchnicki
61. *Computer-Aided Simulation in Railway Dynamics,* by Rao V. Dukkipati and Joseph R. Amyot
62. *Fiber-Reinforced Composites: Materials, Manufacturing, and Design,* P. K. Mallick
63. *Photoelectric Sensors and Controls: Selection and Application,* Scott M. Juds
64. *Finite Element Analysis with Personal Computers,* Edward R. Champion, Jr., and J. Michael Ensminger
65. *Ultrasonics: Fundamentals, Technology, Applications: Second Edition, Revised and Expanded,* Dale Ensminger
66. *Applied Finite Element Modeling: Practical Problem Solving for Engineers,* Jeffrey M. Steele
67. *Measurement and Instrumentation in Engineering: Principles and Basic Laboratory Experiments,* Francis S. Tse and Ivan E. Morse
68. *Centrifugal Pump Clinic: Second Edition, Revised and Expanded,* Igor J. Karassik
69. *Practical Stress Analysis in Engineering Design: Second Edition, Revised and Expanded,* Alexander Blake

70. *An Introduction to the Design and Behavior of Bolted Joints: Second Edition, Revised and Expanded*, John H. Bickford
71. *High Vacuum Technology: A Practical Guide*, Marsbed H. Hablanian
72. *Pressure Sensors: Selection and Application*, Duane Tandeske
73. *Zinc Handbook: Properties, Processing, and Use in Design*, Frank Porter
74. *Thermal Fatigue of Metals*, Andrzej Weronski and Tadeuz Hejwowski
75. *Classical and Modern Mechanisms for Engineers and Inventors*, Preben W. Jensen
76. *Handbook of Electronic Package Design*, edited by Michael Pecht
77. *Shock-Wave and High-Strain-Rate Phenomena in Materials*, edited by Marc A. Meyers, Lawrence E. Murr, and Karl P. Staudhammer
78. *Industrial Refrigeration: Principles, Design and Applications*, P. C. Koelet
79. *Applied Combustion*, Eugene L. Keating
80. *Engine Oils and Automotive Lubrication*, edited by Wilfried J. Bartz
81. *Mechanism Analysis: Simplified and Graphical Techniques, Second Edition, Revised and Expanded*, Lyndon O. Barton
82. *Fundamental Fluid Mechanics for the Practicing Engineer*, James W. Murdock
83. *Fiber-Reinforced Composites: Materials, Manufacturing, and Design, Second Edition, Revised and Expanded*, P. K. Mallick
84. *Numerical Methods for Engineering Applications*, Edward R. Champion, Jr.
85. *Turbomachinery: Basic Theory and Applications, Second Edition, Revised and Expanded*, Earl Logan, Jr.
86. *Vibrations of Shells and Plates: Second Edition, Revised and Expanded*, Werner Soedel
87. *Steam Plant Calculations Manual: Second Edition, Revised and Expanded*, V. Ganapathy
88. *Industrial Noise Control: Fundamentals and Applications, Second Edition, Revised and Expanded*, Lewis H. Bell and Douglas H. Bell
89. *Finite Elements: Their Design and Performance*, Richard H. MacNeal
90. *Mechanical Properties of Polymers and Composites: Second Edition, Revised and Expanded*, Lawrence E. Nielsen and Robert F. Landel
91. *Mechanical Wear Prediction and Prevention*, Raymond G. Bayer
92. *Mechanical Power Transmission Components*, edited by David W. South and Jon R. Mancuso
93. *Handbook of Turbomachinery*, edited by Earl Logan, Jr.

94. *Engineering Documentation Control Practices and Procedures*, Ray E. Monahan
95. *Refractory Linings: Thermomechanical Design and Applications*, Charles A. Schacht

Additional Volumes in Preparation

Introduction to the Design and Behavior of Bolted Joints: Third Edition, Revised and Expanded, John H. Bickford

Geometric Dimensioning and Tolerancing: Applications and Techniques for Use in Design, Manufacturing, and Inspection, James D. Meadows

Computer-Aided Design of Polymer-Matrix Composite Structures, edited by S. V. Hoa

Mechanical Engineering Software

Spring Design with an IBM PC, Al Dietrich

Mechanical Design Failure Analysis: With Failure Analysis System Software for the IBM PC, David G. Ullman

ENGINEERING DOCUMENTATION CONTROL PRACTICES AND PROCEDURES

RAY E. MONAHAN

R. E. Monahan Associates
Fort Myers, Florida

CRC Press is an imprint of the
Taylor & Francis Group, an **informa** business

CRC Press
Taylor & Francis Group
6000 Broken Sound Parkway NW, Suite 300
Boca Raton, FL 33487-2742

First issued in paperback 2019

ISBN-13: 978-0-8247-9574-0 (hbk)
ISBN-13: 978-0-367-40188-7 (pbk)

Visit the Taylor & Francis Web site at
http://www.taylorandfrancis.com

and the CRC Press Web site at
http://www.crcpress.com

To my wife, Elaine, without whose support, inspiration and encouragement I would never have written this book. She is truly my best friend and someone I love very much.

Preface

This book is a very comprehensive discussion of the contents of a good engineering documentation control system for use by technical and manufacturing personnel such that they will be better able to understand and implement good engineering documentation control practices and procedures. Very little information exists today on this subject in our public or private libraries. Specifically, this book is intended for use by design and manufacturing engineers, program managers, documentation control managers, company executives, and anyone interested in managing, developing, or revising an engineering documentation control system. Simplicity and the use of good common sense are stressed throughout the book in all practices, procedures and forms.

This book primarily provides the information for a good engineering documentation control system so that if a company is establishing a new engineering documentation control system or intends to change to an automated system, such as manufacturing resource planning (MRP) and a Computer-Aided-Design (CAD) system, they will have the basics already in place for an effective automated system.

Some of the key items discussed in this book are as follows:

A. Recognize that the term configuration management encompasses the term engineering documentation control.
B. A good configuration planning system should be in place for an effective engineering documentation control program.
C. A configuration identification system should be in place which uses a minimum number of identifiers such as a product number, a part number, a change number, and a serial number (or lot number); and contains a very effective documentation release system in order to control our documentation from the birth to death of a product.
D. A fast and accurate system of change processing is required using a good set of interchangeability rules.
E. A complete analysis of proposed changes is required as to their need, their cost, and the problem solution.
F. A cost analysis of all non-interchangeable changes is required.
G. A good status accounting system of product changes is required in order to insure that the as-built and as-shipped configuration of a product is identical to the as-designed configuration.
H. Any automated configuration management system should incorporate the basics for a good engineering documentation system as described in this book.

The author has prepared this book as a result of doing many seminars and consulting assignments for various-sized companies across the United States during the past twenty years. Also, the author was employed with Control Data Corporation in Minneapolis for over twenty-five years as the Director of their Configuration Management Program.

This book contains primarily the author's experiences and examples that the author has assembled on the subject of engineering documentation control. The author initially received excellent information on the subject of configuration management by attending seminars on the subject at UCLA (University of California at Los Angeles). These seminars, though, were Department of Defense-oriented and hence were not totally applicable in private industry.

The author has taken what he considers the best parts of those early seminars and made them applicable to private industry. As an example of this effort, this book contains many terms and definitions which were initially obtained from those seminars but which are still viable in today's "high-tech" world of business.

A glossary of terms is provided throughout the context of this book as well as in the Appendix. These terms and definitions are vitally needed in a good engineering documentation control program so that everyone speaks the same language.

This book provides the basics for a good engineering documentation system for private industry but does not exclude, of course, companies working under government contracts insofar as their engineering documentation system is essential to their kind of business. Though the material contained in this book is dedicated primarily to the control of hardware products, it is assumed that software products can be handled similarly. Hence, all the basics of a good documentation control system should be applicable to both hardware and software products. More information on the control of software products can be found in many existing Department of Defense standards which are listed in the bibliography of this book and which will assist you in more thoroughly understanding the various elements of configuration management.

Ray E. Monahan

Acknowledgements

The author is indebted to the following people and organizations who have contributed to his knowledge of engineering documentation and configuration control. Oscar O. Akerlund, Richard G. Albers, Donald M. Avedon, Grayme L. Bartuli, Charles D. Herzog, William F. Hyde, Steven Kapernaros, Clifford M. Lieske, Randall M. Omlie, Charles S. Knox, Frank Watts, Gerald I. Williams, Gary D. Wruck, 3M Company, Brisch-Birn and Partners, Ltd., Inc., Cardiac Pacemakers, Ceridian Corporation, Control Data Corporation, Honeywell, International Federation for the Application of Standards (IFAN), Pfizer Medical Systems, Standards Engineering Society (SES), University of Minnesota, University of Wisconsin-Milwaukee.

In addition, the author is deeply indebted to his daughter, Karen Odash, who did the typing for this book, and to Grayme Bartuli for preparing the figures and reviewing material.

Last but not least, the author is very grateful to Marcel Dekker, Inc., for having the confidence in me.

My heartfelt thanks to everyone.

Contents

Chapter 1
Introduction

Companies in private industry struggle to control their engineering documentation, such as their drawings and specifications, which describe the content of a hardware or software product the company is trying to design, manufacture, sell, and maintain. One of the many problems associated with these activities is knowing where to start in developing or revitalizing an engineering documentation control system. This book provides the information necessary to fill the void that private industry experiences, such that they can insure having a good engineering documentation control system.

1.1 Initial Concepts and Definitions

One of the first and most important items in an engineering documentation system is a good dictionary or glossary of terms so that everyone using the system will understand and use the same language. Though this appears to be a very basic requirement, it is one of the major items missing in an engineering documentation control system.

Hence, what follows are the definitions of the first three terms needed in a good engineering documentation control system. Other terms are presented as appropriate throughout the book and are contained in an alphabetical listing in Appendix A3.

1.1.1 Engineering Document

An *engineering document* is a drawing, specification, process, artwork, parts list, bill of material, or other type of document which originates in the engineering department and relates to the design, procurement, manufacture, test or inspection of items or services as specified or directed therein.

The term *engineering document* encompasses all documents as stated in this definition which are generated or initiated in the engineering department and are related to, or are part of, the documentation for the design of a hardware product. We must have a clear definition of this term before we can discuss what makes a good engineering documentation system.

1.1.2 Controlled Engineering Document

A *controlled engineering document* is under document identification control, which consists of the assignment of an initial unique document identifier (part number) imprinted on the document, and revised in accordance with the rules of interchangeability.

It is extremely important that every company has a good set of interchangeability rules for a successful engineering documentation control system. Later on in Chapter 8 we will discuss interchangeability rules in full detail and present a complete set of rules in Appendix A1 so that each company may use them immediately .

1.1.3 Approved Engineering Document

An *approved engineering document* is a controlled engineering document which has been approved by engineering, as a minimum, and by other departmental functions as your company policy may dictate.

It is extremely important to note that the number of approval signatures on any engineering document should be minimal - so what is a minimum number of approval signatures? The originator of the engineering document should be the first approval signature, the second approval signature should be the person responsible for the design of the product, and the third approval signature is the person responsible for the manufacture or producibility of the product. These three signatures should be all that is required on any engineering document. My experience has shown that companies have a range of anywhere from one to ten approval signatures on an engineering document. The smaller the number of approval signatures, the faster the processing of the initial approval of that document and subsequent changes. More than three signatures is too many in this quality-minded age. Indeed, we might have to include a signature from the quality function if that is a major function in your company equal to either engineering or manufacturing.

Since the late 1950's and early 1960's, a term which has been used and is considered very similar to "engineering documentation control" (the subject of this book), is a term used within the federal government, and is called *configuration management.*

1.1.4 Configuration Management

Configuration management is a discipline for providing a systematic or organized approach to planning, identifying, controlling, and accounting for the status of a product's configuration, from its inception throughout its life (that is, from birth to death).

This term is used almost exclusively with companies in the private sector that deal with contracts from agencies of the federal government, such as the Department of Defense. As you can see, the definition of *configuration management* is extremely complete. First, you must have a plan of how you are going to manage the configuration or documentation of your product; you must have a numbering system such that you can use it to identify the content and types of documentation of your product; you must have some kind of a change control system to control changes to your product; and last but not least, you must be able to account for the status of a product's configuration due to changes which may be incorporated in the design for manufacturing in the future.

1.1.5 Configuration

The term *configuration* is defined as the technical description and arrangement of parts or assemblies or any combination of these which are capable of fulfilling the fit, form or functional requirements defined by the applicable product specification and drawings.

Hence, the term configuration is no more than the documentation which defines a new product design. It is in fact the technical description of the arrangement of parts or assemblies which go into making up this new product design. Now, when we look back at the term documentation and the term configuration we can easily see that they are in fact synonymous. From this point on, we will be using the term configuration management with the same meaning as engineering documentation control.

Let's get one thing clear. I am not advocating that every company change the title of their system from engineering documentation control to configuration management, but if I were to start evolving a brand new engineering documentation system, I would use the term *configuration management.*

The other part of the term configuration management which may need to be defined is the term management. I do not care to define this term as each of us has our own definition of management and there are many definitions of this term already. The important thing to remember as we work into a system called *configuration management* is that we are trying to manage the configuration or documentation of a product and we are going to do that by using the proper planning, identification, control and accounting for the status of a product(s) documentation or configuration.

1.2 Reasons for a Configuration Management System

So why should we use a configuration management system to control the engineering documentation of a product? There are several reasons and possibly many more than I will show here, but I would consider the ones listed below the most important.

1. It provides an organized and systematic method for a good engineering documentation control system.
2. It provides all the elements necessary for a good engineering documentation control system.
3. The virtual technical complexity of our products today requires that we have a good system of configuration management.

4. State-of-the-art technology is ever-changing in the world today. We can change the functions of our products by just changing the resistor or transistor on a printed circuit board assembly.
5. The use of both hardware and software in our products requires a good system of configuration management.
6. Complex customer and contractor interrelationships exist in today's business world because of the technical complexity of the products designed and manufactured today.
7. Product liability suits which are being brought out almost every day against companies requires that we have a good system of configuration management.
8. The new ISO 9000 series of standards on quality systems requires that we have a good configuration management system controlling our new products.

1.3 Objectives of a Configuration Management System

If those are the reasons why a good program of configuration management is required, what are the objectives of a good configuration management system? They are as follows:

1. We must be able to plan, identify, control and account for the status of a product's configuration.
2. We should achieve logistic support of a product at the lowest life cycle cost.
3. We do not want to stifle the creativity capabilities of our designers and engineers.
4. We must have an efficient and fast change control processing system.
5. We must have uniform use of all procedures connected with a good configuration management system by *all* company employees.

1.4 History of Configuration Management

We have introduced the term *configuration management*. You may want to know the history behind the development of the term in government and in industry. Please now look at Figure 1.1. You will note that this term evolved in the federal government primarily as we know it today under the Department of Defense. Until the late 1960's, we had many different configuration management programs in the federal government. There was one for the navy, one for the army, one for the air force and so on. In the late 1960's, Secretary of

Thor Program Policy and Procedures	**July 1957 - September 1959**
Atlas Program Policy and Procedures	**July 1959 - December 1961**
Titan and Minuteman Program Policy and Procedures	**July 1961 - Present**
AFSCM 375-1, Configuration Management	**June 1962**
NPC 500-1, Apollo Configuration Management	**May 1964**
AFSCM 375-1, Configuration Management	**June 1964**
AMCR 11-26, Configuration Management	**July 1965**
NAVMATINST 4130.1, Configuration Management	**September 1967**
- - - - - - - - - - - - - - - - -	- - - - - - - - - - - - - - - -
DoD Directive 5010.19, Configuration Management	**July 1968**
DoD Instruction 5010.21, Configuration Management Implementation Guidance	**August 1968**
*** MIL-STD-480, Configuration Control - Engineering Changes, Deviations and Waivers**	**October 1968**
*** MIL-STD-481, Configuration Control - Engineering Changes**	
*** MIL-STD-482, Configuration Status Accounting Data Elements and Related Features**	
MIL-STD-490, Specification Practices	
MIL-S-83490, Specification Acquisition	
- - - - - - - - - - - - - - - - -	- - - - - - - - - - - - - - - -
MIL-STD-973, Configuration Management	**April 1992**

Figure 1.1

History of configuration management. *Standards have been superseded by Mil-Std-973.

Defense Robert B. MacNamara issued an order directing his department to evolve a single system of configuration management for all services to use. In July, 1968, the first Department of Defense Directive 5010.19 entitled "Configuration Management" was issued by the Department of Defense. As shown in Figure 1.1, other DoD documents and Mil standards on the subject and related to the subject of configuration management were also released in that year. In April, 1992, a brand new Mil standard 973 entitled "Configuration Management" was released. This new Mil standard establishes the configuration control requirements, procedures and formats for engineering change proposals, request for deviation/waivers, notice of revisions, specification change notice, and configuration status accounting. In addition, it includes the necessary direction on configuration identification, Class I and Class II changes, baselines, serial numbers, technical reviews and audits. This new Mil standard 973 now supersedes the following existing Mil standards: 480, 481, 482, 483, 1456 and 1521. This new standard is a good reference document for young companies in private industry to have in their libraries.

1.5 Paperless Systems

Many companies are not only struggling with controlling their engineering documentation today but they are also struggling to automate all of their documentation forms and procedures, such as a change order form and the subsequent processing system required for handling changes. It is the belief of most people in the documentation area that before we can have good paperless systems and use of automation we must have all the elements required and a good configuration management system in place. As an example, we should have a minimum number of identifiers in our system such that we only need a product number of some kind, a change number, a serial number, and a good part number system. Any required forms in an automated system should be designed such that they are user friendly; that is, they are easy to use and understand; the simpler the better. Many companies that are handling changes today have change forms which are extremely complex, and are hard to understand and use. That situation must be cleared up before we proceed to automation. (See Figure 18.3 for a description of a totally paperless system and see Appendix A5 for a listing of software packages available today for configuration management automated systems.)

But let's get one thing clear. It is not necessary to have everything for a good configuration management system in place and working manually prior to starting to automate to create a paperless system. It is entirely possible that with the right kind of data base system you could start out initially using some type of automated system which will handle all of the forms and procedures required to identify, control and account for the status of changes in your system.

1.6 Needs for Using Automation

The needs for using automation in an engineering documentation control system are as follows:

1. An automated system will provide you with more timely and accurate information.
2. Data required for managing the configuration of a product is captured at the source rather than elsewhere in your company.
3. Procedures which are required within your configuration management system can be more consistently executed throughout your company.
4. An automated system provides the capability of being able to share the same information among the various departmental functions in your company.

1.7 Summary

This book will provide you with all the elements required for a good engineering documentation control system, or, using the new term just defined, a good configuration management system. It is necessary that we thoroughly understand what these elements are. We have already discussed the point that a good glossary of terms must be available as part of your system. We have introduced a relatively new term to private industry, that is, the term configuration management, which we will be using continually throughout this book. Configuration management encompasses all the elements required to control our engineering documentation associated with the design and manufacture of a new product. But before we embark on the rest of this book we must understand that there are four major sub-topics associated with configuration management. They are configuration planning, configuration identification, configuration control, and last but not least, configuration status accounting for our products. And now, the rest of this book contains the definitions, forms and procedures and some very helpful hints on how to manage the configuration or documentation of your products based on the four major elements of a configuration management system.

Chapter 2
Configuration Planning

Most people in the documentation field wonder why we have to think about planning our documentation when we have all the forms and procedures in place for processing our documentation and controlling it. The reason is that sometimes we have variations which require specific activities to take place as far as different types of documentation that we may need to handle in defining the new product design. For instance, a specific customer may require that you use his part number system and his change forms. How are you going to handle that type of activity without doing some planning ahead of time? That's what is meant by the term configuration planning.

2.1 Definition

Configuration planning is the process of determining how a product shall be configured, documented, and supported.

One of these terms needs to be further defined. We have used the term *product* in the previous chapter and again here in the subject area of configuration planning.

2.2 Product

A *product* is any item produced and offered for sale or lease by your company.

A very simple definition but extremely important for fully understanding how we are to plan and control the documentation for our new product designs.

2.3 Product Design Team or Concurrent Engineering

One of the new concepts that has arisen in the 1980's, which a lot of companies in private industry have incorporated, is the use of a *product design team*. A product design team is usually established in order to increase the interaction and communication between the various departmental functions of a company, such as between engineering and manufacturing. Also, the companies that have used this concept have found that it reduces and minimizes your product development and manufacturing time. Hence, we can get our new products to our customers a lot earlier than before. Another new term used in this same context is *concurrent engineering*. The use of a product design team or the concurrent engineering philosophy is extremely helpful in trying to manage the configuration and documentation of our new products. A product design team can be established for any new hardware or software product development and is usually organized at the conceptual phase of a product. Members of the product design team represent all the major company functional disciplines as required by the activity involved in the development of a product. Some of the disciplines represented on the product design team would be engineering, manufacturing, programming, manufacturing engineering, purchasing, configuration management, production control, quality assurance, publications, and other company functional disciplines which may be required during the activity of the product design team. It has been found that usually the representative of engineering, at least initially, chairs the product design team and as time goes on other functional disciplines may take turns chairing this team, but until the product is fully developed and tested, engineering usually remains as chair of this activity.

It has been well known for many years that a "confrontation" exists between the engineering and manufacturing disciplines in a lot of companies. The use of the product design team concept has reduced this confrontation almost to zero. I know of a large multi-national, multi-divisional company that introduced this design team concept back in the early 1980's, along with the introduction of a new quality discipline and has found many, many company benefits through the use of this concept. Some of these benefits are as follows:

1. Designing, producing, and testing the first product off the production line at almost 100% per its product specification. This has been almost unheard of previously in our design/manufacturing areas.
2. Rather than taking possibly three to five years to develop a product, by using this concept a product can be developed in less than one year.
3. Use of the product design team also has a tendency to reduce the number of future changes which usually are required in a new product design.

Trying to measure these things is hard to do. But if we can reduce our development time from five years to one year, this is a substantial reduction in getting our products to the marketplace. The use of the product design team concept does not replace the authority and responsibility of the project engineer, who has responsibility for a new product design. Engineering still must be responsible for the design of a new product and manufacturing is responsible for the producibility of that new product. Nobody else or no other company function is responsible for those two activities. This principle is violated many times in companies today and once this activity is straightened out, many problems seem to disappear.

2.4 Use of a Baseline System

When we are in the process of designing and manufacturing a new product design, we must have check points along the way to insure that we are properly adhering to the various elements required for a new product design and to insure that we have one element com-

Key: D - Draft A -Approved R - Review / Update	Market Reqmt's	Strategy	Product Reqmt's	Config Mgmt Reqmt's	Product Assurance & Certification	Marketing Plan	Mfg Delivery Plan	Maint Plan
Baseline 1 Definition & Strategy	A	A	D	D	--	--	--	D
Baseline 2 Development	R	R	A	A	D	D	D	A
Baseline 3 Design and Verification Testing	R	R	R	R	A	A	A	R
Baseline 4 Product Delivery & Enhancement	--	--	--	R	R	R	R	--
Baseline 5 Phaseout	--	--	--	--	--	R	R	--

Figure 2.1
Example of a product baseline review system.

pleted either in series or parallel before we go on to the next element. Figure 2.1 shows the structure of a typical product baseline review system. Note that the items across the top of this diagram are items required in trying to manage a new product development. As an example, we must understand what the market requirements are before we develop a new product design and what is the company strategy for this design? What are the product requirements? What are the special, if any, documentation requirements needed? How are we going to handle the product assurance and certification of this product in case we need a UL (Underwriters Laboratories) listing or a CSA (Canadian Standards Association) listing? How are we going to handle the impact of any future changes to this product and its certification? What is our marketing plan for this particular product and when do we intend to try to deliver initial units to our customers? And what about maintaining this product for our customer? Are we a company that is so large we might need our own maintenance organization and do our own product maintenance? Or are we going to have a second party or third party do our maintenance for us? Or are we going to tell our customers to send there product back to the factory if they have any problems? What is our maintenance plan? These are all questions that need to be answered as we go into the development of a new product design.

The baselines that we should establish are shown on the left hand side of the diagram in Figure 2.1. Baseline number 1 consists of defining the product and determining the strategy for this product. In addition, we must have certain items completed before we can move on from baseline 1 to baseline 2, such as the market requirements must be defined and the overall company strategy for this new product design must be defined. After we have those two items completed then we can move onto baseline 2, which is a further development of this product. Now we have to define the product requirements and the special documentation requirements for this product and we must start looking at the product certification requirements and the marketing plan as well as the overall manufacturing plan.

When we have this activity completed, we can move on to try to accomplish baseline 3, which has to do with the design and verification testing. Here we have to make sure we have completed and know the product certification plans, we must know the marketing plan and also the intent for the manufacturing and delivery of this new product design.

Baseline 4 has to do with the final product delivery and any future enhancements which might be made to this new product design as shown in the chart in Figure 2.1. Last but not least, we finally get to a point in the overall control of this product concerning phase out. Are we going to have a subsequent brand new product design to replace this product? Are we going to revise this product when this product reaches maturity? These are all questions which must be answered in the overall development of a product and we in the documentation area must be very much concerned about the answers to each of these activities.

The baseline chart in Figure 2.1 is only intended as a guide. You may want to use different tools, or you may want to do things differently, or you may want to alter this chart. That's fine, as this is only a guide in the overall development of your product. The important point is that each company should have some kind of a product baseline system, and whether they use the term baseline or phases or some other term makes absolutely no difference. The important point is that you should have some kind of check point system to guide you in the overall design, development, manufacture and maintenance of a new product design to insure that your documentation is compatible with the overall company product strategy and goals.

2.5 Product Life Cycle

Every new product design must cover a complete life cycle as we described in the definition of configuration management. We are trying to plan, identify, control and account for the configuration status of a new product from the time it is conceived throughout its whole life, that is, from birth to death. The chart shown in Figure 2.2 is a brief attempt to illustrate what the major responsible functions may be in the overall control of a new product design and how they relate to the life cycle of a new product and all of the various activities that occur during this overall time frame. So when we look at Figure 2.2, we see that there are four major phases, or we may call them baselines, which a product goes through in its overall design, development and manufacture.

First of all, we have the conceptual phase which gets us into the area of trying to define the overall design requirements for this new product in the form of a product specification, which comes under the functional definition of this product. A product specification is one of the most important documents required in the documentation of a new product design. But in my experience in talking with many companies, I find that at least half do not have a product specification which controls the overall functional and performance requirements for a product. How else can we control these items if we don't have some form of a product specification which is subject to overall change control, just like engineering drawings and other types of specifications? We will discuss this subject more in Chapter 6. The other major item under the functional definition is an item called bill of material. The bill of material is also one of your most important documents in con-

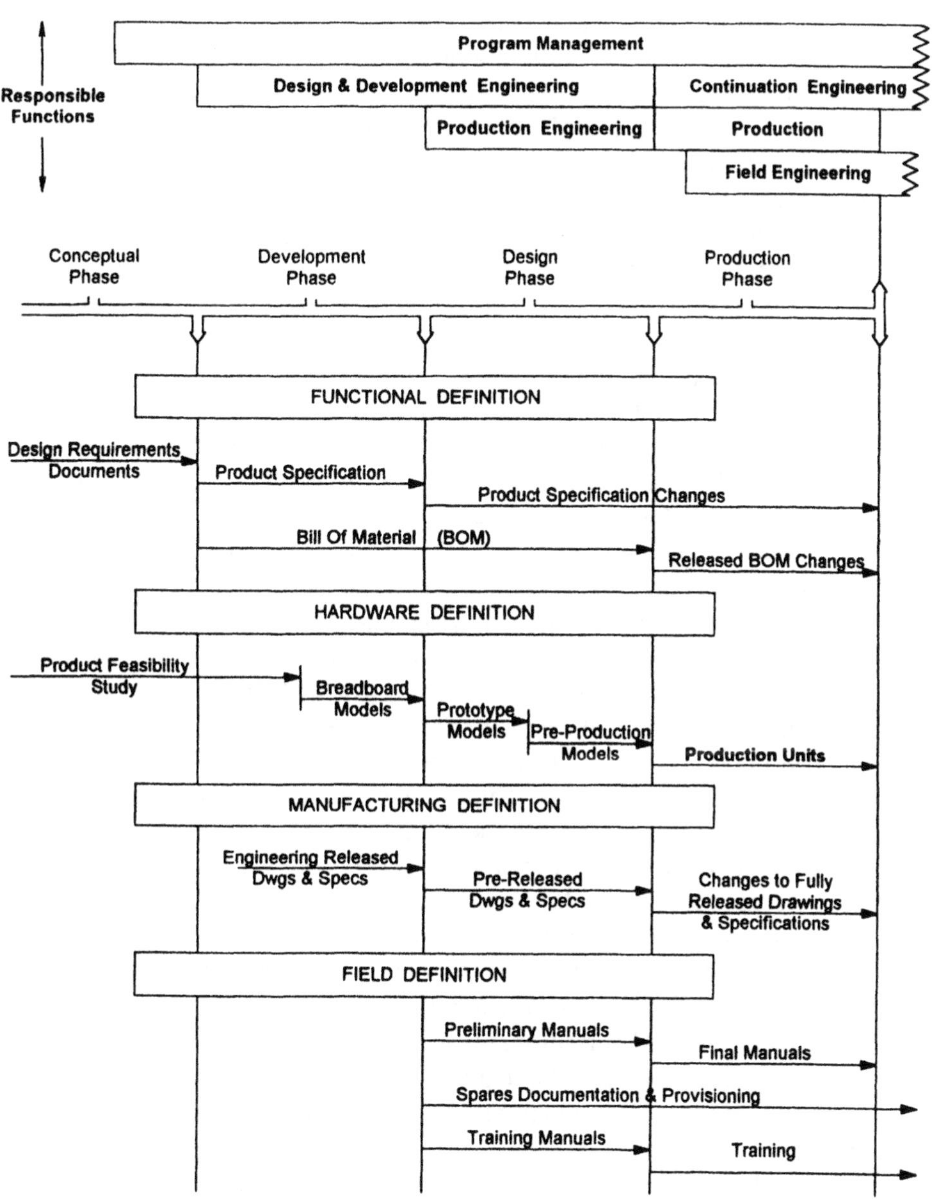

Figure 2.2
Product life cycle.

trolling the overall configuration and documentation of a product. We will discuss this term in much more detail in Chapter 4.

The hardware definition of a product comes about in five different forms. Traditionally, we have product feasibility models which we will develop to prove or disprove new product design ideas. From that, we get into breadboard models to further prove out these new ideas or concepts. From breadboard models, we develop prototype models in which we build a complete prototype of the product to prove out the product design. From the prototype model, we get into the pre-production models, which are models generally built to prove the produceability of this particular product. And last but not least, we get into the actual building of production units of this new product design. All of these different types of models or units come under the heading of defining the hardware definition for this product design. These four different hardware models can be easily executed using computer simulation and modeling.

Next, we have the manufacturing definition, which consists of the various drawings and specifications which document all of the elements, including piece parts and assemblies, for example, of a new product design. These drawings and specifications must be subject to some kind of a document release system which should exist in your company. Typically, a document release system consists of an initial engineering release wherein we do not have much control over these types of drawings and specs. Then we get into a pre-release mode as shown in Figure 2.2. Pre-released drawings and specifications are those in which our risk and confidence level is not very good. It certainly is possible in the design phases shown in Figure 2.2 that we could have a mix of engineering-released drawings and specifications as well as pre-released documents as production begins on a given product. We will discuss release in much more detail in Chapter 5.

Under the field definition, we discuss the various kinds of manuals and spares which may be required for your product. Manuals can be of many forms. There may be a manual consisting of many pages or only an instruction sheet but, in any case, these manuals must be controlled documents and they may come initially in the form of a preliminary manual and then finally they are fully released, making them a final manual. These manuals or instruction sheets are usually generated to define the operation and maintenance instructions required for a given product. In addition to manuals, we have to be concerned about the possible spares which are needed to service a new product, how we are going to support the new product with spares, who is responsible for the spares inventory and how are we going to document spares. Spares documentation and provisioning is an extremely important part of the overall documentation for a new product. And of course last but not least, we must also be concerned about training the people using our products, if necessary, such that they can fully understand how to operate and maintain our product most effectively.

2.6 Configuration Management Plan

A configuration management plan is a document recommended in the Department of Defense system for each and every new product design they acquire. Department of Defense Standard 1456a entitled "Configuration Plan" describes the ingredients for a configuration management plan in the DoD system of configuration management. For purposes of companies in private industry, such a plan might include any or all of the items shown in Figure 2.3. A configuration plan which might be evolved for companies in private industry would include at least some of the items shown in Figure 2.3. For example, you may be doing a special product design for a special customer or customers and you may be required to use their change system and their forms. If that's true, your configuration management plan should indicate this type of activity, including how to obtain a copy of the change order system used by that company. You also might be required to use the customer's part numbers rather than yours. That customer part number system should be defined in your configuration plan and any of the other items shown in Figure 2.3 could be a part of this plan. This plan would be your guide to let everybody know what kind of a configuration management system we are going to use on this special product design.

Another important item that might be included in the configuration management plan is regarding what type of MRP (manufacturing resource planning) system you will be using to manage this product, and if it is a system that is used by people in the total company. It usually is operated within the manufacturing function of the company and the system does include your bill of material. But any major functional area in the company should have access to this system in order to accomplish their part of the input required for the selected MRP system. The question here is which MRP system do we use? Ours or the customers? Many MRP systems are offered on the market today and it is very difficult to say which should assist you in controlling your overall configuration management system. We have listed some of the existing MRP systems in Appendix A5 and the people responsible for their development and sales. A given MRP system may be right for your company but may not be the correct one for another company, so your selection depends entirely on your kind of business.

Product Name	
Development Group Responsible	
Change Orders	**(Use yours or customers)**
Engineering Drawings	**(Use yours or customers)**
Engineering Specifications	**(Use yours or customers)**
Tracking Number to be used	**(Product No. Or P/N or ...?)**
Will there be Field Modifications	**(Do we need a Field Change System)**
Identification Plates	**(Product, UL, CSA, VDE, Warning Labels, etc.)**
Part Numbers	**(Use yours or customers)**
Publications	**(How to Control ?)**
Warranty & Recall Procedure	**(Needed ?)**
Serial Numbers	**(Use yours or customers)**
Automation Considerations	**(Use yours or customers)**
Top Level Assembly	**(Drawing Needed ?)**
Approval of CM Plan	**(Highest Level Possible - President !)**

Figure 2.3
Items for consideration in a configuration management plan.

2.7 Summary

In order to have an effective configuration management system some type of planning has to take place in order to best accomplish the documentation of a new product design. Each company should have some kind of a product baseline control system as described in Figure 2.1 and a configuration or documentation plan, if required, in order to properly execute the product objectives for a new product. Some type of chart which defines a product's life cycle, such as the one shown in Figure 2.2, is a very effective tool in training your new engineers, designers, technicians, manufacturing people and customers on the importance of the design and manufacture of a new product.

Therefore, it is extremely important that this subject be addressed by the product design team in their initial discussions regarding designing, manufacturing and maintaining a new product design.

Chapter 3
Ingredients for a Configuration Management (Engineering Documentation Control) System

The ingredients for a configuration management system are as shown in Figure 3.1, which depicts all or most of the ingredients which would be a part of any good configuration management system. A company in private industry may have other ingredients in their own configuration management system for their kind of business, but the items in Figure 3.1 can be considered a minimum for most companies. Let us review this chart briefly because this will be our road map for the balance of this book.

Configuration management, as previously defined, consists of the necessary planning, the identification program, the controlling of changes to a product and the accounting of a product's configuration from birth to death. As you can see in Figure 3.1, there are several

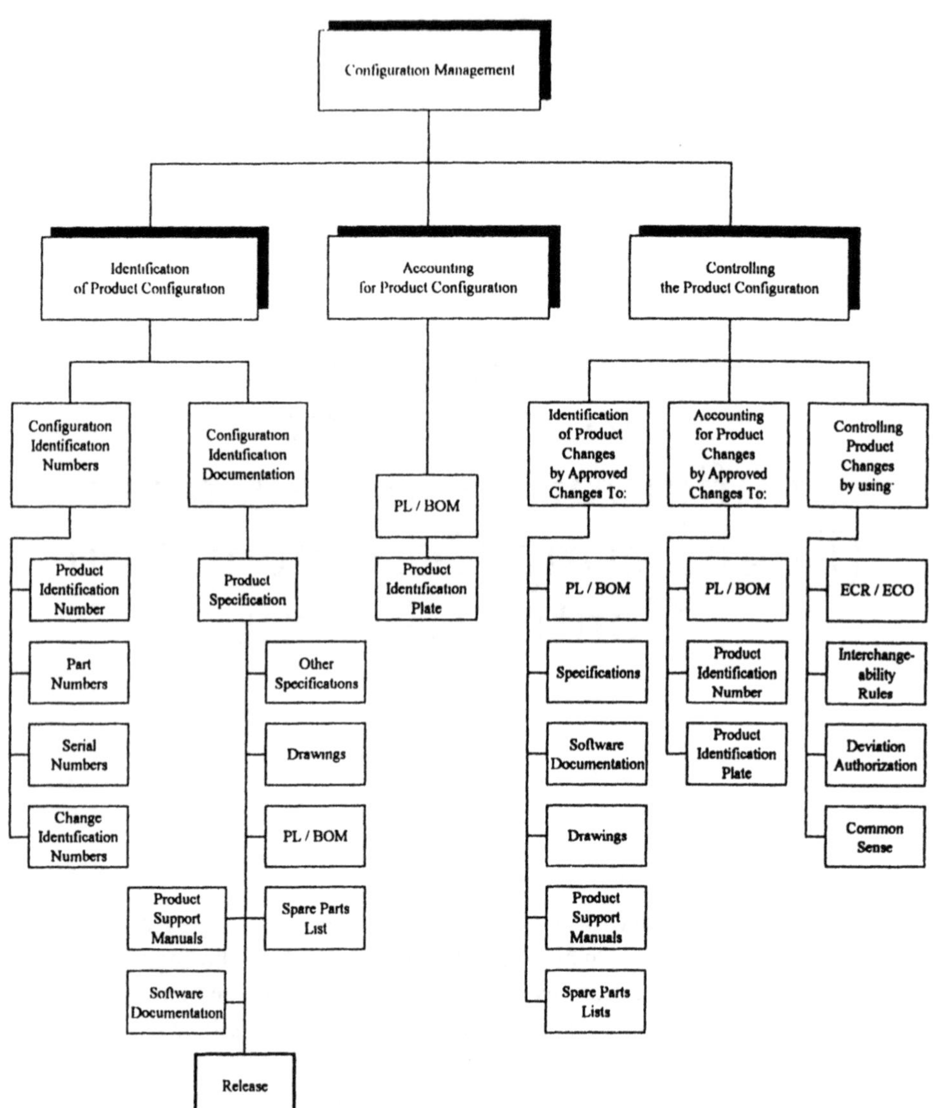

Figure 3.1
Ingredients of configuration management.

sub-headings under each of these major areas. The chart does not show the major element of configuration planning as we have already discussed that area and the value of such planning in controlling our documentation. This chart depicts the other three major elements of configuration management, namely, identification, accounting and control of a product's configuration.

3.1 Configuration Identification

Identification of a product's configuration consists of the various number systems required to control and identify the documentation. Under configuration identification, we show the types of documentation required to control and identify our product's configuration. This documentation, of course, would be identified by the numbers shown under the block entitled configuration identification numbers. The major piece of documentation required under configuration identification documentation is the product specification, which we have discussed to some degree already. Other documentation includes the various drawings, specifications, the bill of material and any parts lists for various assemblies, any software documentation, any spare parts list, and any manuals. The heart of any good configuration identification documentation system is the subject of release, which we will discuss in Chapter 5.

3.2 Configuration Status Accounting

Under accounting for a product's configuration, the chart in Figure 3.1 shows the bill of material (BOM) as the key item in accounting for the status of a product's configuration and hence, a bill of material must be kept up-to-date at all times to show the various parts and assemblies and documentation, if necessary, which are required to design, manufacture and maintain a given product.

3.3 Configuration Control

Under the control of a product's configuration, this element breaks down again into three major areas:

a. Identifying any changes to the product by changes to its documentation, such as specifications, bill of material, drawings, manuals, software and spare parts list. Your company may have additional types of documentation that should be shown in this area, but this chart was developed to be as generic as possible in order to fit the needs of most companies. If you were to redraw this chart, you would include any special documentation requirements for your kind of business.
b. The second area under controlling a product's configuration is being able to account for product changes by approved changes to the product's bill of material, the product identification number (see Section 4.1), the product identification plate, and possibly a product configuration log (see Section 13.4). Again your company may have other documents which would replace or add to these documents.
c. Under the area of configuration control of a product, we have the controlling of product changes by the use of the following items:

(1) Some kind of a change request and order form which may be called an ECR (engineering change request) and an ECO (engineering change order) and a FCO (field change order), if control of field changes is required (see Chapter 10) by your company.
(2) A good set of interchangeability rules are needed.
(3) Some kind of a deviation or waiver form might be required for your system (see Section 7.5).
(4) Last but not least, we should all be using good common sense when we are trying to control changes to our product.

The chart in Figure 3.1 is a good summation of the items required in a good configuration management or engineering document control system. Whether your company has each of the items shown in this chart, less or more items, or different titles for some of the items listed would be entirely up to your company and its kind of business.

I would like to go back now to the configuration identification numbers on the chart in Figure 3.1. Under the term configuration identification numbers, we have an item at the very bottom that says cage code. This is a federal code which is used by the federal government to identify the various manufacturing facilities in private industry that deal with the various agencies of the federal government. The acronym "CAGE" means "commercial and government entity code" and if you do not have such a code, it is available through the Defense Logistics Center in Battle Creek, Michigan. The CAGE codes are also recorded in *Federal Handbook H4-1*.

3.4 Summary

Figure 3.1 is the road map, that we will be following throughout the rest of this book to explain the major and minor elements of a good configuration management or engineering documentation control system. Each company should have a similar chart to depict the ingredients of their engineering documentation system so that all employees can be trained in your system.

Chapter 4
Configuration Identification

This major element of a configuration management system consists of the various numbering systems used in managing our documentation. In addition, it encompasses all of the different types of documentation which we use in order to define the various elements of our new product design. Configuration identification is defined as follows:

Configuration identification is the technical documentation which is properly identified and defines the approved configuration of products under design, development, and test, in production or in the field.

As stated previously, *configuration identification* includes identifying by some number system the various kinds of documentation we use to define the various parts, assemblies, etc. of a new product design. Any good *configuration identification* system should contain the following major requirements:

1. It should be simply developed such that it is usable and understandable by all employees of your company so that they can communicate with the various numbering systems effectively.
2. It should be contemporary so that it can handle the state-of-the-art technology of a new product design.

3. It should be flexible such that it can adapt to many different situations.

We will now define and illustrate the various kinds of numbers that should be used, at least as a minimum, in a good configuration identification system.

4.1 Product or Model Identification Number

A product or model identification number is either a significant or non-significant number which identifies a specific product and is usually assigned and controlled by a company's marketing or sales organization. Sometimes, though, only product names are used for identifying a given product or model. If we look at some of the appliances in our home today, for instance a microwave, you will find a product identification plate on the back. The product identification plate typically includes a product or model number (these terms seem to be used interchangeably within industry) and also, a serial number and possibly a top level assembly part number. The product or model number is used primarily when we have a problem with a device and we have to obtain some new parts or spare parts in order to repair the device. This number is also used by sales people in the actual sale of your product and you will usually find it on the associated sales documentation.

Now, should you use the term product or model number? A product number is usually some kind of alpha numeric or straight numeric number which identifies a specific product or group of products. Whereas a model number usually is used to identify a revised version of that product. You might have a product number that is 6632 and the model number would be subsequent to the product number, such as 6632-A or 6632-1 or 6632-100. There is apparently no standard in industry on this subject and I have only tried to provide a guideline on this kind of a numbering system for your products. Though a name, as I said previously, will work on your product identification plate or in discussing it with your customers, it is possible that the name is significantly long and will be hard to use in the documentation which describes this particular product. Hence, a product or model number or both is usually used to identify a new product. Also, it should be known that product numbers are usually used to identify products throughout their entire life cycle.

4.2 Serial Numbers

Serial numbers are usually assigned to the various units which are produced on a manufacturing line of a given product. They are usually non-significant but they can be significant. They are usually all numeric and are sequentially assigned, usually starting with number 101. Serial numbers 1 through 100 are usually only used to identify prototype or preproduction models. They do identify a specific unit and are used for tracking that unit of a given product when it is on the manufacturing assembly line or after it has come off the manufacturing assembly line for future change activity. In addition, the serial number is used for designating what the effect of a change is on the manufacturing line. With a given device or product which is physically very small, it is possible that serial numbers cannot be used. They can then be controlled by some kind of a date activity scheme or by a group or lot number for those built in a given quantity. For instance, normally we can say that we have a product XYZ in which we have built 2,000 units. These 2,000 units would then be assigned serial numbers, possibly starting with 101 on up through 2100. Each unit coming off the assembly line would then be given its appropriate serial number usually assigned by the production control function of your company. But if it is a very small product, such as a monitor for a television set, you may have to use a lot number for a group of products to specifically identify them.

In using a new serial number for implementing changes made to a product, it is entirely possible that we may incorporate the change at serial number 1252 and other new units after that. Hence, we use the serial number for effectivity of the change. This would mean that a particular change installed in serial number 1252 and on would incorporate the change and that the change would not have been incorporated in units prior to serial number 1252.

Some companies use block point serialization. This type of serialization means that you establish a given serial number in manufacturing as a block point some time in the future, at which time you will incorporate not only one change but possibly several changes at the same time. So all of those changes would have an effectivity at the same serial number 1252, as we used in the previous example. This is a good method for installing a group of changes on your manufacturing assembly line.

Once a given product is delivered to a customer, you will know which serial number that customer has. Should that customer have a problem with his product, they can relay the information on that product to the factory by using the product number and serial number. Again, my experience has shown that the use of serial numbers is extremely important for a good configuration identification and change control system.

Serial numbers are usually assigned at the end of the production line by the production control department, but if its more appropriate for your company because of your kind of business to assign them at the beginning of the production line, there is simply nothing wrong with doing that. The advantage of assigning serial numbers at the end of the production line is that sometimes units do not come off the production line in sequence and hence, your serial numbers could get mixed up. But there are advantages and disadvantages to both systems depending on your company's requirements.

4.3 Change Numbers

A change number is the number that appears on the change form used to track changes to the product. It is usually a non-significant number that is all numeric and is assigned sequentially. Like part numbers, change numbers can be assigned by a data base. The change number appears in the revision block of an engineering drawing. It is sometimes shown in a parts list or bill of material and it is usually shown in a configuration log or list that is developed by the configuration management organization for tracking changes. A typical configuration log would include the engineering change number as well as the date assigned and what products are affected (a "where used" file). It is the third number required in the overall configuration identification system and these numbers are controlled and assigned by the configuration management organization in your company.

4.4 Part and Document Numbers

4.4.1 System Requirements

The requirements for a good system of part and document numbers are as follows:

1. Keep your system as simple as possible.
2. Part numbers can either be non-significant, such as 12345678, with no significance attached to any of the digits, or the part number may be significant, such as 12AB6792, wherein significance is placed on all or some of the digits in the part number. Please see Figure 4.1 for the advantages and disadvantages of significant versus non-significant part numbers. Generally in industry, my experience has shown that about half the companies use non-significant part numbers and half use significant part numbers. There is no strong feeling one way or the other. I prefer a non-significant number because there is a longer life and less error in a non-significant number than there is to a significant number. Typically, companies run out of numbers in certain categories of a significant number. Also, a non-significant part number is more cost effective to use than a significant part number. See Figure 4.2 for examples of different company part number systems.

   When establishing or converting from a manual to an automated documentation system, it will not make any difference whether or not a significant or non-significant part number system is used.
3. Revision letters are not a part of the part number but are subsequent to the part number. They are used to indicate interchangeable changes to a part number, such as letters A, B, C, D, E, etc. Revision letters I, O, Q, S, X and Z should not be used per Mil-Standard-100. Assuming the rules of interchangeability are adhered to, parts should only be stocked or inventoried by part number and not include the revision letter status. You should be able to use a revision A part or a revision F part out of inventory.
4. A part number must include the document number, as shown in Figure 4.2.
5. Use a central assignment function for controlling all part and document numbers. Assign one part and document number at a time

so that you don't draw any blanks in your part number system. A part and document number log should exist such that part numbers are assigned in sequential numbering order. In today's world of automation, part and document numbers can be assigned through use of a data base. The log should include the following items:

- Product number (or numbers) used-on.
- Engineer responsible for the design.

Significant	**Nonsignificant**
versus	
Example: BAAX2-2AB-2	**Example: 409872**
Limited Life	Unlimited Life
Requires strict control	Very Flexible (Less errors)
Must publish data books	Very simple for users
Security breached	Better security
Not readily understood (misinterpretations)	Eliminates interpretation
Length can vary	Uniform length
Higher maintenance costs	Most economical
Describes document or part	No significance
Categories can break down	No categories

Figure 4.1

Significant vs. non-significant part and document numbers.

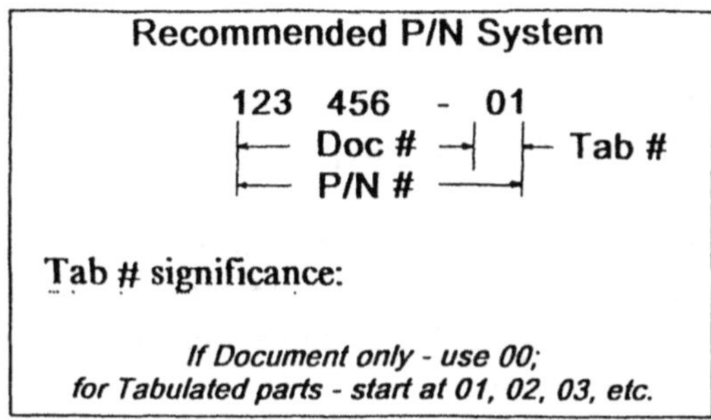

General Motors
Example:
123 456 7

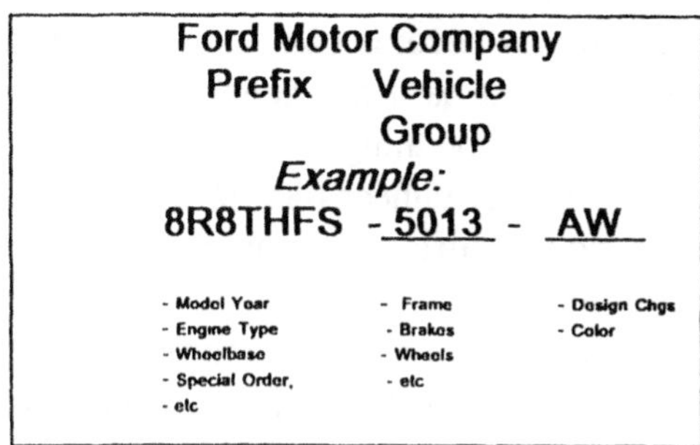

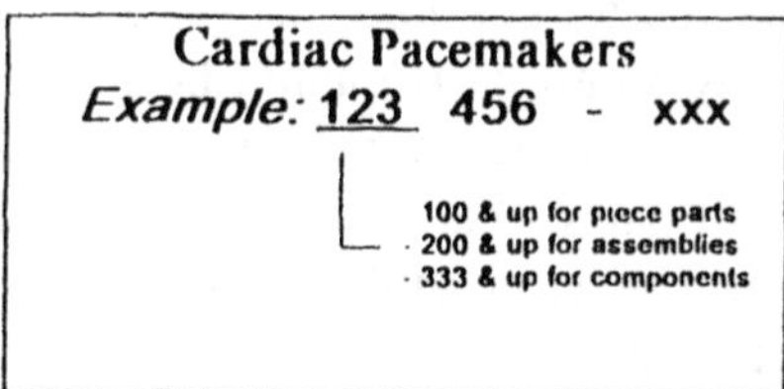

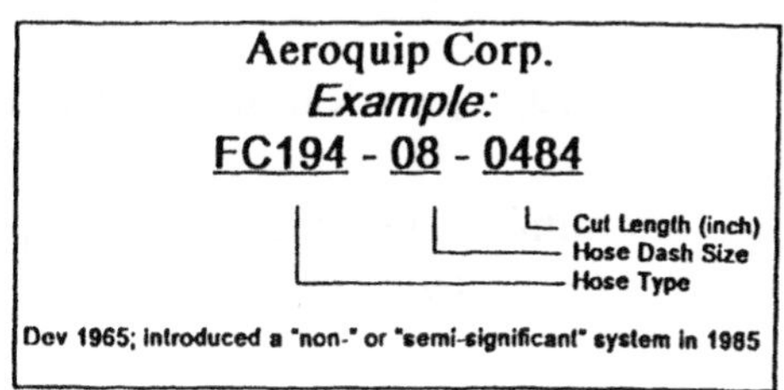

Figure 4.2

Part and document numbers. Note: Spaces in part number examples for legibility only.

- The date the number is assigned and any other information that might be appropriate for your company's operation.

6. A separate numbering system for sketches is not needed as the existing part and document number system can work for sketches as well as regular drawings. This is an example of keeping your system as simple as possible.

   Typically, some companies use a sketch numbering system such as SK001 for the first sketch, SK002 for the second sketch, and so on. This system only becomes an extraneous and unneeded system as eventually you will assign a part number to it and then you have two numbering systems to track. Some people may argue that the sketch may become unused or obsolete and you are losing a part number. But, is it not better to have one rather than two identifiers for the same thing!
7. Keep it free from any unnecessary changes. This is one of the reasons that a non-significant number is better than a significant number system.
8. Make sure your system has long range flexibility such that it can be used for identifying many different kinds of documentation.
9. Part Marking. Part numbers are used to mark parts which are spared or need to be tracked for possible changes and are state-of-the-art type parts. Many small parts cannot be tracked by part number because of their size, hence, sometimes a date code or lot number is used for small parts. Part marking is very expensive and should be avoided whenever possible and as noted previously, you never mark a part with the document revision level subsequent to the part number.

4.4.2 Part Number System Usage

Your part number system should be capable of being *used* for any or all of the following items:

1. Assembly drawings.
2. Detailed part drawings.
3. Installation drawings.

4. Product support manuals should be controlled by your part number system as they are subject to changes in your change control system. A part number should be assigned to each of the operating and maintenance instruction sheets or manuals associated with your product.
5. Software can use numbers out of your part number system because we want to be able to manage our software just like our hardware.
6. Specifications should use numbers from your part number system as it is absolutely unnecessary to create a new system of numbering for specifications. Specifications are under engineering change control and we do not want to confuse people with more than one numbering system.
7. Specification control drawings should use your part number system.
8. Use your part number system for test equipment, parts and assemblies documentation.
9. Tool and gage drawings should use the part number system. Assign a block of your part numbers to the tool design department which they can use. Now, whether they should choose to use your engineering change control system or their own is something that has to be decided between the engineering and tool groups. It can work both ways, but there is no need to create a separate numbering system for tool and gage drawings. Remember, tools and gages might be sold with the product. Hence, another good reason for using the company part number system.

4.5 New Part Number Release Cost

Figure 4.3 is an example of the possible cost which can be incurred over the life cycle of a part due to releasing a new part number to identify a particular part. As shown in Figure 4.3, just about every possible activity which can happen against this particular part during its entire life cycle is accounted for. There are costs accumulated for part qualification, certification, receiving inspection, documentation release, change control, approved vendor list, purchasing activity in storage and in inventory, and the cost of design, manufacturing and the on-going continuation engineering services. As shown in this particular example, done for a large multi-division, multi-national, multi-billion dollar company, the total cost exceeds five thousand dollars for just one part number. It is important to remember that this is just an example. The cost in your company would depend on the size of your company and the volume of your sales activity. Your cost could be considerably more or less for releasing a new part number. The important thing to remember here is that you should know what it costs to release a new part number in your company and use the chart in Figure 4.3 as a guide to assist you in collecting these costs.

4.6 Documentation Requirements

In Figure 4.4 we have depicted in a block diagram the various items included in the family tree structure of a product's top level assembly. As we review the items in Figure 4.4, we will discuss some of the major elements of this typical family tree.

4.6.1 Bill of Material and/or Parts List

A bill of material is a separate document which carries the same part number as the part number of the top level assembly drawing and should never be shown on the drawing itself, just as a parts list for a sub-assembly should not be shown on the sub-assembly drawing itself. In this age of automation, having a clerical person refer to an assembly drawing for the various parts and sub-assemblies which make up a given product is a very cumbersome and tedious process. So, it is inherent that we keep the bill of material and/or parts list separate from the assembly type documentation. The only exception to this rule might be in the case of a weldment assembly drawing or in new releases of CAD drawings, which show the parts list on the drawing, but this data is ASCII code data and is electronically available to be integrated with the MRP system.

Typically, I have found in industry that parts list and bill of material are not well defined and hence you will find some people who use them almost interchangeably when the real fact is that they are as different as night and day. First, let's look at the definitions of these two terms.

AREA	AVERAGE COMPONENT COST EXAMPLE	BACKGROUND
Qualification, Failure Analysis and Certification	$1,000.00	3 million annual budget. Qualification cost per component ranges from $0 to $25,000
Receiving Inspection and QA	$1,660.00	10 million annual budget for common parts. 50% avoidable costs.
Documentation		
Release	$ 60.00	300 average document cost
ECO's, etc.	200.00	4 ECO's per document at $250 per ECO.
QVL	40.00	5 Tabulations per drawing
Purchasing	$1,425.00	$32.33 weighted average purchase order cost. 350,000 purchase orders per year or eight purchase orders per year x 5.5 years average life = 44 purchase orders per life of the average component. 1 of 3 items purchased is a common part.
SUBTOTAL	$4,385.00	
		Other Assumptions
other		- 3,000 new common part P/N's per year
		- Constant release rate.
Storage	$ 100.00	
Inventory	50.00	
Engineering	180.00	
Manufacturing	150.00	
Engineering Services	150.00	
Cataloging/Retrieval	100.00	
SUBTOTAL ESTIMATE	$ 730.00	
TOTAL	$5,115.00	

Figure 4.3
New part number release cost.

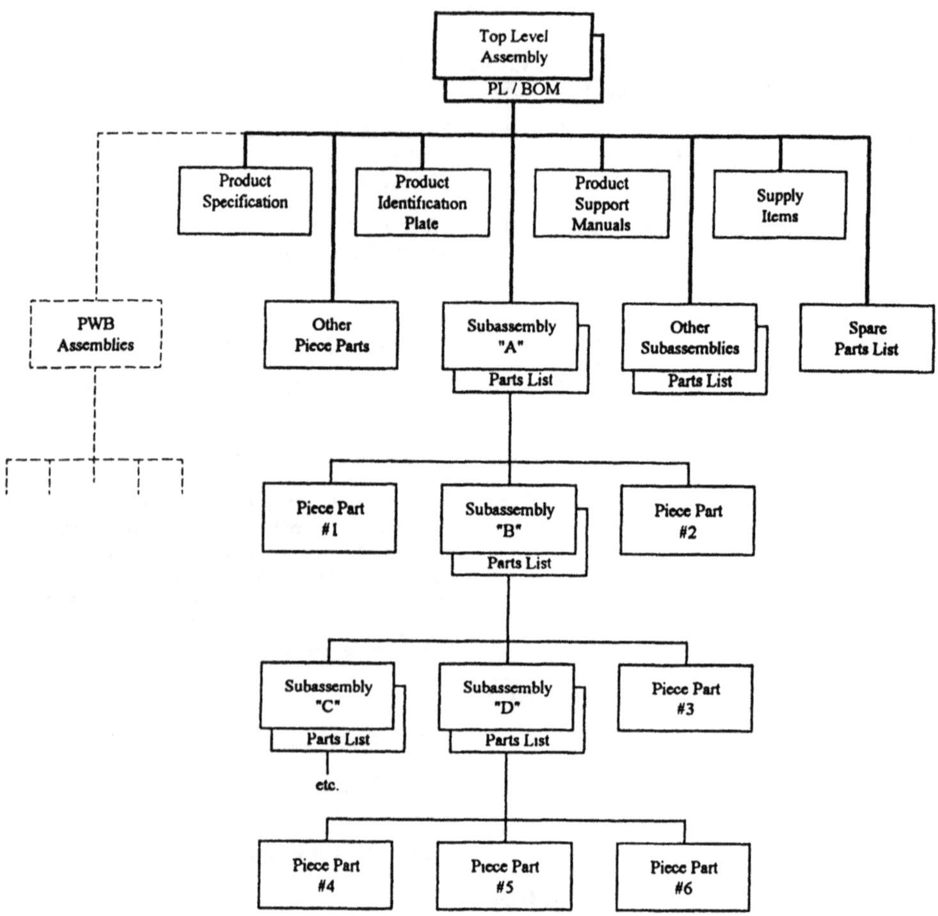

Figure 4.4
Typical product documentation requirements (family tree).

Parts List. A listing of only those items needed to build a specific assembly in the family tree or bill of material of a product.

Bill of Material. A complete listing of all piece parts, assemblies and any other items required to build and ship a product to a customer. A bill of material encompasses all parts lists required in a product plus any other required items for shipping the product, such as product support manuals.

To illustrate the difference between a parts list and a bill of material let us look at Figure 4.4. The parts list for the top level assembly drawing would only include those items shown at the first level of blocks immediately underneath the top level assembly. For instance, the parts list would include sub-assembly A and its parts list but it would not include sub-assemblies B, C, D in the block diagram. The parts list would only show such items as the piece parts and the assemblies required in the final assembly of this product to make up the top level assembly. Other items that might be included on the parts list: the product specification, the number of the spare parts list if there is one, the numbers of the operating and maintenance manuals, and any supply items, such as magnetic tape or disks, that might be required for a disk drive which is a product.

On the other hand, a bill of material will include each and every item at every level, as shown in the block diagram in Figure 4.4 directly under the top level assembly. It will include all of the sub-assemblies and their part numbers and all of the piece parts that are required for this particular product, right down to the last nut, bolt, screw, transistor, resistor, integrated circuit, etc.

Also, note on the block diagram that an assembly block is shown for PWB assemblies which becomes a phantom assembly. PWB is the acronym for printed wiring board. The idea here is to try to get printed wiring board assemblies at the highest level possible in the family tree of a product so that if changes need to be made to this kind of an assembly, and if they are non-interchangeable changes which require a part number change, that part numbers only need to be tumbled from the block that has the PWB assemblies printed in it and possibly the top level assembly part number. Hence, only two levels would have to change, rather than having the PWB assemblies scattered as sub-assemblies in lower parts of the family tree structure in which case we would have to tumble a lot of part numbers to get to the highest part number affected, namely, the top level assembly part number. This is a kind of "trick of the trade." This trick can be used also with any other kind of common item or common assembly item that you might have in all of your products. It isn't required that you do it just for printed wiring board assemblies.

So, what kinds of items can go into a bill of material as shown in the block diagram in Figure 4.4? Specifications, the product identification plate (which is a label), the operating and maintenance manuals, the packaging material that is used for packaging your product for shipment, and any other labels, such as a label for UL (Underwriters Laboratories) or a label for CSA (Canadian Standards Association), or any other safety labels which might be required. Other process consumables, such as soldering material or supply items, magnetic tape or computer disk, also might be a part of your bill of material as well as any additional literature which may be required for shipment. Sometimes tools, gages, and fixtures have to be a part of the shipment of a given product and if so they would also be listed in the bill of material.

So how do we know what items should be a part of bill of material? Your data base, depending on how it's structured and on what MRP II (material requirement planning) system is used, will help to define which items must go into a bill of material. But the most important thing, within the company is that *common agreement* must occur on what does or does not go into a bill of material. As long as all the functional disciplines of your company (i.e., the product design team) concur on what goes into the bill of material, this activity should really eliminate any problem which may occur in the future on the content of the bill of material. You must be consistent on your bills of material from product to product so that you don't have to have a separate set of rules for every product or product family created within your company. Only *one* bill of material should exist per product.

A bill of material is extremely critical to the overall control and process of identifying, controlling and accounting for the status of a product's configuration from birth to death. It is the key document which controls the configuration or documentation of a product. A bill of material should only contain a listing of all the items that make up that product per your company rules, by part number and not by part number and revision letter. If revision letters are used in a bill of material, your bill of material will be subject to many unnecessary changes just to update the revision letters. It is an excessive and unnecessary expense to your company to have revision letters shown in the bill of material. This is assuming that your company is adhering to the rules of interchangeability.

For just-in-time manufacturing, it is imperative that you only have a single level bill of material structure such that you can accomplish the requirements of just-in-time manufacturing. One of the common industry problems for bills of material is that a bill of material has so many levels, such as shown in the previous block diagram in Figure 4.4, that when the bill of material printout is run on an MRP system, it takes an entire weekend. Hence, it is important that we try to eliminate levels of assembly within a product. We should also be trying to minimize the number of cost centers, when using an MRP system, in switching over to a just-in-time manufacturing capability. The use of a phantom assembly as shown in Figure 4.4 is a way of trying to eliminate levels of assembly.

The items that can be spared in your bill of material should have a column in the bill of material identifying which items should be spared such that if a printout is required for the bill of material, it can be done to identify all items which are considered to be spare parts.

4.7 Component Documentation

Documentation for components is a very important part of controlling engineering documentation. The definition of a component is as follows:

A *component* is a purchased part, assembly or product which is purchased from a outside vendor. Examples of components are fasteners, resistors, integrated circuits, clamps, power supplies, etc.

Typically, components are documented in the form of a specification control drawing defined as follows:

A *specification control drawing* is a document which specifies the needed requirements of a component for your company such that it can be purchased from an outside vendor.

Figure 4.5 is an example of a specification control drawing developed by a small company in Minneapolis. This is a drawing of a slotted pan head machine screw which is normally purchased from an outside vendor. Only those dimensions and specifications are shown which are needed to be controlled by your company. You do not have to show each and every dimension of a part or component on a specification control drawing. Only those items and specifications which

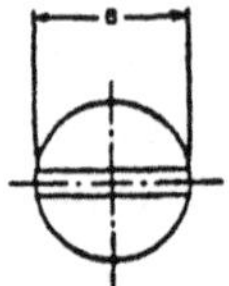

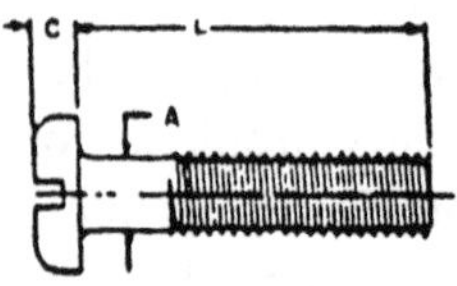

		A		B		C	
Nom. Size	Threads Per Inch	Body Diameter		Head Diameter		Head Height	
		Max.	Min.	Max.	Min.	Max.	Min.
2	56	.0860	.0717	.167	.155	.053	.045
4	40	.1120	.0925	.219	.205	.068	.058
6	32	.1380	.1141	.270	.256	.082	.072
8	32	.1640	.1399	.322	.306	.096	.085
10	24	.1900	.1586	.373	.357	.110	.099
10	32	.1900	.1658	.373	.357	.110	.099
.250	20	.2500	.2127	.492	.473	.144	.130

PART NUMBER LISTING BY LENGTH AND BODY SIZE

L Length	2-56	4-40	6-32	8-32	10-24	10-32	.250-20
	810201						
.125	-009	-018	-030				
.187	-010	-019	-031	-044			
.250	-011	-020	-032	-045	-061	-077	
.312	-012	-021	-033	-046	-062	-078	-093
.375	-013	-022	-034	-047	-063	-079	-094
.500	-014	-023	-035	-048	-064	-080	-095
.625	-015	-024	-036	-049	-065	-081	-096
.750	-016	-025	-037	-050	-066	-082	-097
.875	-017	-026	-038	-051	-067	-083	-098
1.000		-027	-039	-052	-068	-084	-099
1.250		-028	-040	-053	-069	-085	-100
1.500		-029	-041	-054	-070	-086	-001
1.750			-042	-055	-071	-087	-002
2.000			-043	-056	-072	-088	-003

MATERIAL - STEEL SAE 1010-1020	FINISH - CAD or ZN

APPLICATION - Slotted head screws should be used where ease of adjustment or tightening is of primary concern. For normal fastening application, the cross recess drive should be used.

SPEC. CONTROL DWG.

MATERIAL NOTED			COMPANY NAME & LOCATION		
FINISH NOTED			TITLE SCREW-MACH, PAN HEAD SLOTTED		
A	1/8/94	REL'D	AMS	DWN AIDB	SCALE NONE	DWG. NO 810201-TAB	REV A
REV	DATE	ECO NO.	APVD	CHKD OCK	DATE 12/30/93	SHEET 1 OF 1	
REVISIONS				APVD AMS			

Figure 4.5
Example of a specification control drawing.

you desire to be controlled and reviewed by your receiving/inspection department need be shown.

4.8 Classification and Coding System

There are a number of classification and coding systems which exist in this country today. Some of these systems are available for purchase from an outside vendor or are developed internally by the company itself. It has been found in industry that those who try to invent their own classification and coding system are usually doomed to failure. Typically, systems invented internally do not encompass all of the needs and requirements for your kind of business. Hence, I would recommend an organization that developed what I consider one of the best classification and coding systems in industry. The Brisch-Birn and Partners, Ltd., Incorporated, located in Ft. Lauderdale, Florida. This company over many years has developed a very intelligent classification and coding system which when developed for other companies works extremely well. Figure 4.6 shows the wheel chart developed by Brisch-Birn and Partners, Ltd., Inc., it is possible to classify and code any piece part or assembly you desire. The code turns out to be a five digit numeric code, very easy to use not only for sorting identical parts and similar parts for inventory purposes but great for use in the group technology environment of manufacturing. There are many other types of classification and coding systems on the market today, most of which are spin-offs of this particular system. If you are interested in this kind of activity for your company, contact Brisch-Birn and Partners, Ltd., Inc., in Ft. Lauderdale, Florida.

4.9 Use of CAD-Generated Documents

As most of us know, the acronym CAD means computer aided design, which uses automation equipment to generate the necessary engineering documents for portraying the elements of our new product design. There are many different types of CAD equipment on the market today which are used as workstations to generate these documents. The information that I have presented regarding controlling engineering documents, drawings, and specifications can also be used at a CAD workstation. The control of these documents is almost

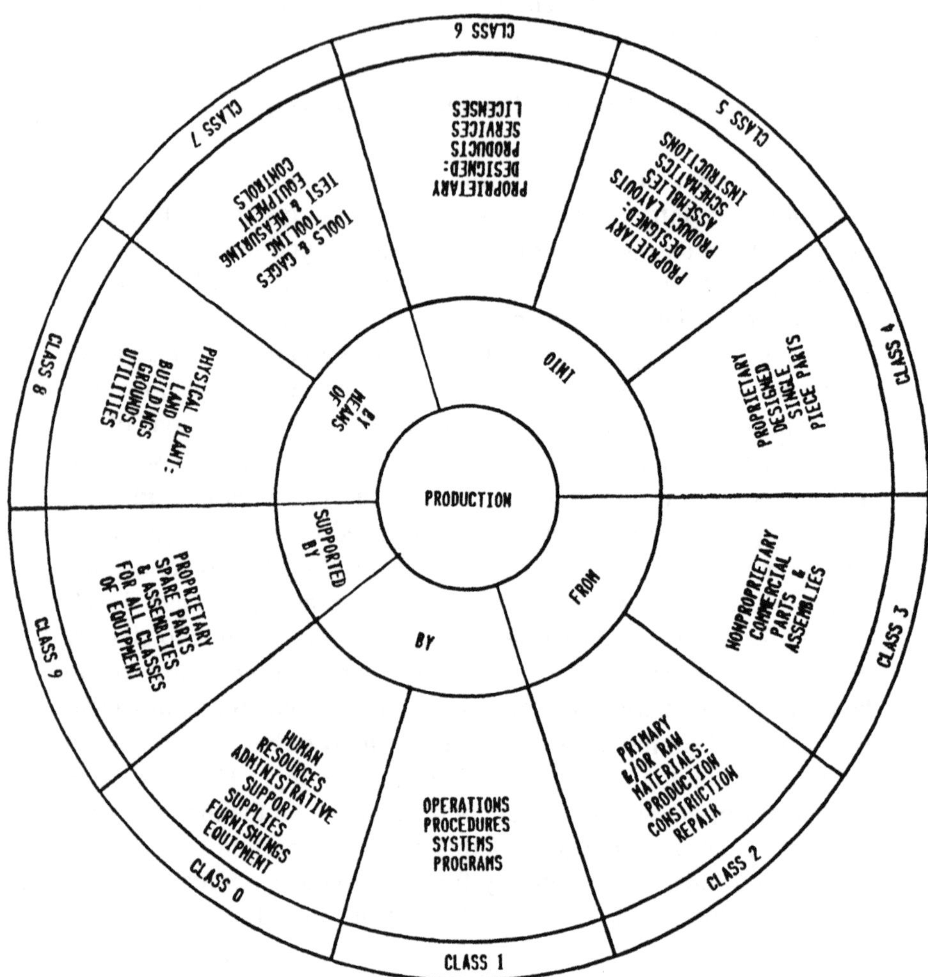

Figure 4.6

Example of a classification and coding system . Typical classification schema of an engineered design products manufacturing firm. This chart is owned and developed by Brisch-Birn and Partners, Ltd., Inc., P.O. Box 22490, 1656 SE 10th Terrace, Ft. Lauderdale, Florida, 33316: (305) 525-3166.

identical to controlling manually generated documents, such as drawings and specifications. A title block would exist on the drawing format which is used in the CAD system. A part number would be assigned to that document, and any change activity to that document would get recorded in the revision block of that drawing. All of the things necessary to portray the various parts and assemblies which make up a new product design can be controlled in almost the identical way in a CAD workstation, which is where the document is generated electronically.

Today, most companies working with CAD-generated documents are not having any problem in controlling them. The biggest problem appears to be in obtaining the needed review and approval of these CAD-generated documents and doing it *electronically*. Though some companies claim to have an electronic review and approval system in place, an outsider's review of these systems usually shows that some parts of their system are still done with paper and handled manually. There has been only one system that I know of today that is what I would call a totally paperless system and yet even this system has a manual "glitch" in it. As Don Avedon, a document management systems consultant, stated recently "we are only 10% paperless today in private industry and if we look out five years from now we will still only be about 15% paperless." Hence, in summary, it will be years before private industry can consider themselves as being in a "paperless society."

4.10 Summary

This chapter has included all of the requirements needed for a good configuration identification system for your company. We have discussed the minimum types of numbers which are required in such a system they include a product model identification number, a serial number, a change number, and a part and document number. We have also identified those items which can be described under a single part number system such that we do not have to have extraneous types of number systems in our overall configuration identification system. We can use part numbers to identify specifications, tool and gage drawings, operating and maintenance manuals and so on, and by doing this we can we keep our configuration identification system as simple as possible. We have discussed the various types of documentation you may have in your particular system but these types of documents are not intended to be all inclusive. Certainly the types of documentation for your company will depend on the nature and kind of business . We have tried to be as generic as possible to assist as many companies as possible that might be evolving or revitalizing their overall configuration identification system.

We have discussed the differences between a bill of material and a parts list and the need for both as well as the importance of having a bill of material which must be kept up to date at all times, as it portrays the current configuration status of a product and its documentation. We discussed the term component and its kind of documentation in the form of a specification control drawing and from there went on to discuss an example of a classification and coding system which might assist you in eliminating identical or similar parts in your system as well as assist you with the use of group technology in manufacturing.

We have noted that when generating CAD or any other computerized engineering documents, the documentation controls used are the same as for manually generated documents.

Chapter 5
Engineering Documentation Release

At the heart of any good engineering documentation control system is a system for releasing documentation from engineering to manufacturing. Though many different types of release systems exist within companies in the private sector of industry, it is difficult to say that one system or another is better than a previous one or adjoining one. Hence, it is difficult to recommend a specific documentation release system that can possibly be good or generic for all companies. But what we can do is try to establish some of the basic terms and definitions for a system that might work for you by using an example.

5.1 Release System Example

This chapter will define an example of a release system which contains what are considered the basics for releasing documentation in a good engineering documentation control system. This system is working for several companies today and it may possibly work for you. So let's get on with describing this example of a release system.

5.1.1 Release Definition

First of all, we must define what we mean by the term release as follows:

Release is the act of supplying valid and fully approved and documented technical information on engineering designed products for both procurement and manufacturing.

5.1.2 Types of Release

Release applies to all technical documentation, such as drawings, specifications, manuals, software documents, processes, tooling, and change documents. Under our example of a release system, there are three different types of release. Namely, engineering release, pre-release, and full release. Each of these types of release may be defined as follows:

Engineering release is the use of a controlled engineering document which is needed for limited procurement of parts and assemblies on breadboard or prototype models.

Pre-release is the use of a controlled engineering document which authorizes manufacturing to proceed with production of a specified number of units, called pre-production models, provided these engineering documents have also been approved at least by the engineering and manufacturing functions.

Full release is the use of fully approved engineering documents in the procurement and manufacture of production models.

5.1.3 Hardware Model Types

Within these definitions we have described four different kinds of hardware models: breadboard model, prototype model, pre-production model and production model. Now let us define these four different kinds of hardware models.*

---

*These models are frequently replaced, in today's world, with computer simulation and modeling done in 3-D at a CAD workstation.

A *breadboard model* is the assembly of hardware representative of a product or a portion thereof used for experimental and testing purposes as well as proving out a new product design function by engineering.

A *prototype model* is the complete assembly of a product used for design testing in determining the adequacy of the total design. This product is built under the direction and control of engineering and not manufacturing.

A *pre-production model* is the complete assembly of a product by manufacturing and not engineering, which is built from at least pre-released engineering documentation during an initial production or pre-production phase to check out production techniques. During this phase, production tooling is fabricated, processes are developed and checked, acceptance tests are run, and final product refinements are made in the design.

A *production model* is a product which is built on the production line from fully approved and released documentation.

In this example, we have identified three types of release and four hardware model types and previously we have defined three engineering document types. Now, let us look at Figure 5.1, which shows the relationship of these items and how release status can be indicated on your engineering documents. First of all, the three release types can possibly be better identified for communication purposes by using Class C for engineering release, Class B for pre-release and Class A for full-release. For change control purposes, it is readily seen in Figure 5.1 that there is no formal change control system which could exist for a Class C type document. A modified or accelerated change control system is probably used for Class B with limited approval by engineering and manufacturing but a change control board consisting of product design team members could be used, and then a Class A document would be under full change control in whatever that system is. The release status is usually shown in the revision block of our drawings. Figure 5.2 shows an example of what the revision block might look like. As you can see, revisions, which are made when a document is in a Class C status, are done by changing the date in the revision block of the drawing and giving a short description of the change. A Class B drawing is noted in the revision block and is given a pre-release status by showing a numeric revision

Type of Release / Document Class	Definition	Under Change Control	Used On	Release Status Indicated By	Subsequent Revisions Indicated By
Full Release / Class A	Fully approved Engineering document, by engineering and other departments as required	Yes	Production and Pre-production Models	Releasing Document at Rev. A & Inserting "Class A Released" in revision block	Change in alpha revisions. (i.e. A to B, B to C, etc.)
Pre-Release / Class B	Engineering controlled document which is at least approved by the responsible engineer and is under an accelerated method of change control	Yes, but no Change Control Board (CCB) is used.	Pre-production and prototype models	Releasing document at Rev 1 & Inserting "Class B Pre-Released" in revision block	Change in numeric revisions. (i.e. 01 to 02, 02 to 03, etc.)
Engineering Release / Class C	Engineering controlled document which is not approved nor under ECO control. Parts procurement requires Engineering Manager approval.	No	Prototype and Breadboard Models	Insert "Class C Unreleased" in revision block	Changing date in the title block

Figure 5.1

Interrelationship of product document classes, hardware models, and release types.

LTR	ECO	DESCRIPTION	DATE	APPROVED
-		Released - Class C	8-8-91	R. Albers
-		Tolerance Updated	9-9-91	R. Albers
01	44985	Pre-released - Class B	9-9-91	R. Monahan
02	45039	See ECO	1-3-92	G. Bartuli
A	44678	Released - Class A	3-7-92	CCB
B	45111	See ECO	4-1-92	F. Watts
C	45222	See ECO	4-9-92	R. Monahan

Figure 5.2
Sample document revision block. ECO approval "signatures" should be written (signed) unless electronic approval is used. Do not use I, O, Q, S, X, and Z in a revision letter system.

of 01 with a description that says "Class B pre-released" plus the date and approval initials. Numeric revisions are used only for Class B type drawings. When we have a fully released or Class A type drawing the document is released as a Class A drawing as shown in Figure 5.2 at revision A, and then the first actual revision to the drawing is revision B, C, and so forth.

It is not necessary for a given engineering document to have to go through all three classes of release. If your confidence factor on a given item is such that you may be taking little or no risk on releasing a document initially as Class A, that is certainly a preferred way to go. If you desire to release a given new engineering document at Class B to start with rather than Class C, that also can be done; in other words, it is not a rule that you must subject each and every engineering document for a product to go through a three-step release cycle. If your confidence and risk factor is such that you can release at Class A initially, that is certainly the preferred way to go. It should be noted that when your document is in a Class A or full release status and letter revisions are used, that the following six characters should not be used in a revision letter system: I, O, Q, S, X and Z. These letters when handwritten can be confused with other letters or numeric characters in your system.

5.2 Ground Rules for Release

It is always good to have a set of rules when discussing the subject of releasing engineering documentation. The following seven rules should govern just about any type of release system:

1. All critical components in a product shall be qualified before full release.
2. Full release of a product can only be effected after satisfying the requirements of a design baseline at which time all configuration-defining documents for that product are released.
3. At any time prior to completion of the product design, a part may be released to manufacturing or to procurement, especially long lead time items.
4. No document should be released unless all supporting and associated documents are also released, except for product support manuals which are difficult to release at the same time.
5. If a product is to be produced and delivered to a customer prior to full release, a deviation should be written and approved stating these facts for quality control purposes.
6. When applicable, all standard products should be approved and listed by a certified agency, such as Underwriters Laboratories (UL), prior to full release.
7. Products should only be shipped to a customer when the documentation has been fully released (or Class A).

5.3 Release Form

Some companies use some type of an engineering or design release form. It is the belief of the author that such a form is unnecessary if it is possible to show the documentation release status in the bill of material which is associated with your product. If a column can be reserved in the bill of material and designated as release status, you can have the release status shown for each engineering document which makes up your new product. Hence, if you use the release system described here, you could use the letters A, B and C to indicate what the current status of each document is in the bill of material and change their status as time goes on in the design and development of that product.

Another method which may be used to record the release status is to create what I call a block-type change control document. This is done by using a change form from your change control system and assigning a number to that change from your change numbering system which will be used in the revision status block of the engineering documents for a specific product. The engineering change form under this description would then list all of the engineering documents required to build and maintain your product and the release status would be shown for each and every document on that change form. This is more of a manual system and hence an automated bill of material system is highly recommended. Hence, those are two methods used to track the revision status of your engineering documents for a given product.

5.4 Summary

There is no doubt that a good release system is required in your engineering documentation control system so that we know what the release status is of any and all documents for a product at any given time. Whether we use an automated system through the use of a bill of material or some type of a manual form it is entirely up to the discretion of your company and its capabilities and resources.

The system described here is only an example of a release system several companies are using in private industry today which really does work with this system and we can track the release status of any document for a new product design.

Chapter 6
Configuration and Change Control Philosophy

This is another one of the four major elements of a good configuration management system. *Configuration control* is defined as follows:

Configuration control is a systematic or organized method for identifying, controlling, and accounting for the status of changes which affect the configuration or documentation of a product.

Hence the three major subheadings under *configuration control*, as shown in Figure 3.1, are as follows:

1. Identify product changes by approved changes to your document and subsequently your parts and assemblies, which make up your particular hardware product. The kinds of documentation impacted usually by a change are your drawings, specifications of all types, bill of material and parts lists, product support manuals, spare parts list, and any accompanying software documentation. In addition, other kinds of documents may be subject to change which are used in your kind of business.

2. Accounting for the status of product changes by approved changes to the bill of material and/or parts list, and possibly the product identification number, the identification plate and a document called a product configuration log, which we will discuss later in Chapter 13.
3. Controlling product changes by using some type of a change control form, a good set of interchangeability rules, some type of a deviation/waiver process and last but not least, by just using some good old *common sense*.

Many of us have had two major problems associated with changes:

Figure 6.1 shows a Volkswagen which has been inundated by so many changes that the documentation outweighs the automobile itself.

In Figure 6.2, we have an example of the same Volkswagen in which the hardware has actually been changed but again, there have been so many changes to the automobile we have literally lost control because now we have no place for the driver to sit and parts of the engine are sticking outside the trunk of the automobile. It is imperative when we start talking about controlling changes that we must be able to identify, control and account for the status of *all* changes to the configuration or documentation of a product.

6.1 Philosophy of Change

A hardware product change affects or could affect many different items associated with that product. A change could have a tremendous impact on the product and cost if we do not realize the *total* impact of that change as shown in Figure 6.3. Some of the items that could be impacted by a given change, which is usually a non-interchangeable change, are specifications, drawings, manufacturing processes, product support manuals, parts on order, parts on hand, parts in process, product certification, bill of material, product safety, software, parts interchangeability, and so on. There are many, many different items which could possibly be impacted by a change and one of the glaring defects in most of the change control systems which exist in our companies today is that they do not cover the process

Figure 6.1
Example of a product overwhelmed with changes and paper.

Figure 6.2
Example of a product overwhelmed with hardware changes. To control changes: identify, control, account.

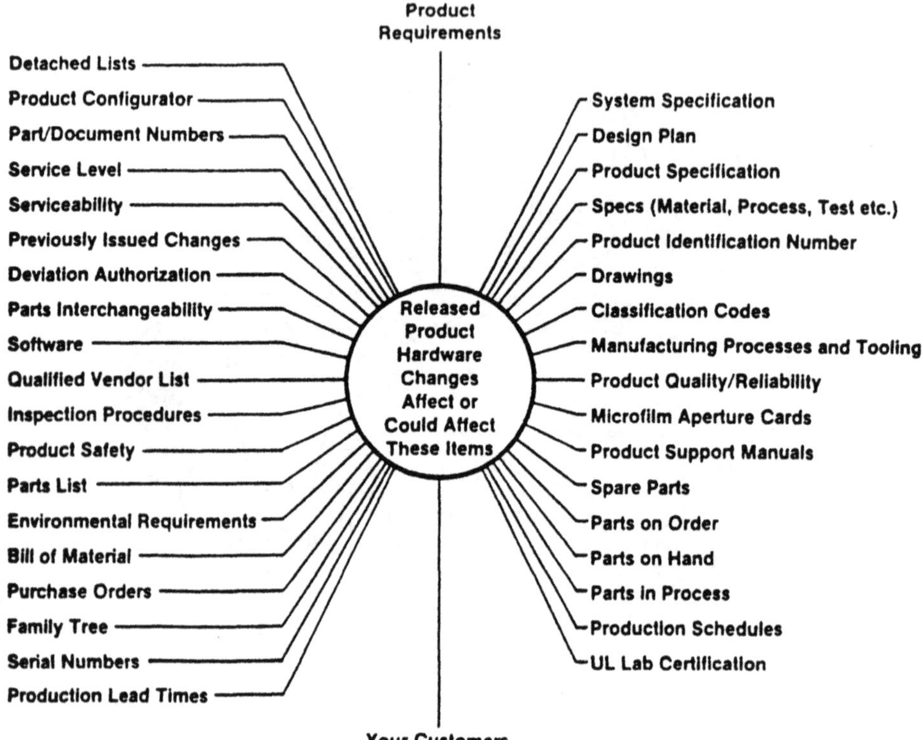

Figure 6.3
Released hardware product change impact.

of obtaining the *total impact of a change*, which is most important if we are going to control changes in our change processing systems. We always seem to have the time to change things but never the time to design it right in the first place.

Though none of us like to see changes to our products, we must recognize the following facts:

1. Changes are a fact of life.
2. Changes must be controlled.
3. Changes are necessary for technological progress.

We will never get rid of change. We can have a moratorium on changes for a given period of time in order to reduce the number of changes which have occurred. But, as just stated, we can never get rid of change if we are going to keep our products current with the state-of-the-art and the use of new technology.

6.2 What Is a Change? What Do We Mean by a Change?

A *change* is a modification to a product which is necessary to meet the requirements of the functional specification, to meet safety standards, or to reduce manufacturing or maintenance costs.

This is a very important definition in your overall configuration control system. Please note that the definition of change says it must *meet* the product functional specification, that is, it must meet all of the physical and functional requirements of this product. We must have a document in our overall engineering documentation control system which is entitled product specification. My experience again has shown that over approximately one-half of the companies who have attended my seminars do not have a product specification. I asked myself, then, how can they manage change in their company and do a total impact of a change if they have no control over the functional and physical characteristics associated with their product. A simple definition for what is meant by a product specification is as follows:

A *product specification* is a document which describes what a product is and what it does by defining its physical and functional characteristics.

It is imperative, if we are going to have total configuration control in our engineering documentation control system, that a product specification *must* exist to control its physical and functional characteristics.

6.3 Managing Change

If we are to try and manage changes in our products, we must meet the following four essential items as listed below:

1. All requirements of the product must be positively identified including the product specification.
2. We must have strict control over all changes.
3. We must maintain a detailed change accounting system to include the date of change authorization, who made the first change, when and where the first change was incorporated in manufacturing, and maintaining revision control of our documentation when incorporating changes.
4. We must recognize that all changes are mandatory for implementation into our product and its documentation when approved and released no matter how small or how large the change may be.

6.4 Change Evaluation

Change evaluation is a very important part of your overall change control processing system. Your change evaluation process should include the following items:

1. Assess the total impact of a change as to its need, its cost, and the problem's solution feasibility.
2. The impact on the product specification.
3. The impact on the product's safety certification, whether it be listed by Underwriters Laboratories, Canadian Standards Association, the British Standards Institute, or any other certified agencies throughout the world.
4. All changes must maintain or improve the product's design integrity.
5. Assess the impact on the interchangeability of parts and/or assemblies.
6. Assess the impact on the product support manuals, such as the operating and maintenance instructions which may be required for your product.

Change evaluation is an extremely important part of the overall change control process. Not taking the time to assess the total impact of a change can result in your company having many unnecessary changes to process in your system. "Doing it right the first time" or "pay me now or pay me later" applies here.

6.5 Summary

The configuration and change control philosophies which have just been discussed are extremely important when structuring or revitalizing a change processing system. But not only for that purpose. If you are considering having some type of an automated change system, you will find it extremely helpful to adhere to the philosophies just discussed.

Chapter 7
Engineering Change Control Forms

The information which should be contained in control forms and the types of change control forms is what we will discuss here, because that is most important. Many companies have change control forms which are totally inadequate or fail in their primary purpose of processing a change. We must understand the total impact of the change, as to its need, cost and feasibility to the solution of the problem.

7.1 Engineering Change Request (ECR) Form

An *engineering change request form* is no more than a form, available to any employee, used to describe a proposed change or problem which may exist in a given product.

Some companies have an *engineering change request form* in their system and others do not. What I have usually found is that those companies considered high production, building about 100 or more products per day or per week, have a change request form.

Those building less than that number do not have a change request form. The person completing this form may or may not know any of the part numbers involved, let alone knowing the product number. They are not necessarily technically trained, so they may not give a recommended solution to the problem. What we usually ask is that the people completing this form fill out as much as they can.

Generally, an engineering change request form will contain the following information:

1. A space for numbering the engineering change request, called an ECR number. ECR forms are typically pre-numbered so that the request can be tracked from the originator through the entire engineering change system.
2. A list of the part or document numbers impacted by this change, if known .
3. State the complete description of the change or problem which has occurred (a picture is worth a thousand words), so if a copy of a drawing or specification can be marked up and attached to the request, this method is much preferred.
4. If known, a recommended solution should be stated on the engineering change request form.
5. Name of requestor.
6. Name of requestor's supervisor.
7. Date of the request.

These are the important items which should be contained in an ECR form. A sample engineering change request form is shown in Figure 7.1. This form is only an example of how an engineering change request form may appear.

An ECR form, when used, should be forwarded to the organization responsible for processing changes in your company. This form, like the trigger of a gun, will start the process of determining the total impact of this change and its best possible solution. Companies not requiring or using an ECR form will usually do one of two things. First, they may combine the ECR form with the ECO form, which we will discuss later and secondly, they may not have an ECR form at all and they will use the engineering change form for requesting changes, or possibly memos are used to request a change.

<table>
<tr><td colspan="5">Engineering Change Request</td><td colspan="2" rowspan="2">ECR No.</td></tr>
<tr><td colspan="3">Product Name:</td><td colspan="2">Affected Product Model Nos.</td></tr>
<tr><td colspan="5">Drawing or Document Numbers Involved (If known):</td><td>Page:</td><td>Of:</td></tr>
<tr><td colspan="7">Reason For Change:</td></tr>
<tr><td colspan="7">Description Of Change or Problem:
(Attach any Documentation that you might have)</td></tr>
<tr><td colspan="7">Proposed Solution:</td></tr>
<tr><td colspan="3">Requested ECO Priority:</td><td>Emergency ☐</td><td>Urgent ☐</td><td colspan="2">Routine ☐</td></tr>
<tr><td colspan="7">ORGINATOR - DO NOT WRITE BELOW THIS LINE</td></tr>
<tr><td colspan="7">ECR Disposition</td></tr>
<tr><td>Assigned ECO No:</td><td colspan="6"></td></tr>
<tr><td>Originator:</td><td>Date:</td><td>Location:</td><td colspan="2">Supervisor:</td><td colspan="2">Date:</td></tr>
</table>

Figure 7.1
Recommended engineering change request (ECR) form.

7.2 Engineering Change Order (ECO)

The titles used for identifying the change control form to process engineering changes vary from company to company but some of them are engineering change notice (ECN), engineering notice, engineering order, and so on. The author prefers engineering change order as it is much more direct, as the definition implies. This form is defined as follows:

An *engineering change order* form is a document which describes an approved engineering change to a product and is the *authority or directive* to implement the change into the product and its documentation during the manufacturing process.

Engineering change order (sometimes known as an ECO) is preferred over other terms because the form, as described in the definition, is a directive for people to do something in implementing this change. Someone has to make the changes to the document or documents affected, others may have to order new parts or revisions to old parts, and assembly people will be responsible for implementing the change on the assembly line. Engineering change order is a very direct term instructing people to do something. It is much firmer than other terms used in industry today, which in some cases take the form or title of a memo. The ECO form shown in Figure 7.2 can be used either manually or in an automated format. The most important aspect of this form is that it includes the minimum amount of information required on any change control form used for processing changes. So let us review these items:

1. Change number. A change number should be assigned to each change order developed. This number is usually all numeric and might start, as an example, at 001 and continue on sequentially. The revision block shown in the change number area in Figure 7.2 is used to indicate any letter revisions to the change order caused by a change in the effectivity status of the change after its approval. *No other changes or revisions should ever be made to the engineering change order form.* If a mistake is made on the change order form other than effectivity, a new change order must

LOGO

Engineering Change Order

DIV | ECO NO | REV

DATE APPROVED:

PAGE 1 OF

OLD PRODUCT ID

NEW PRODUCT ID

CLASS OF CHANGE
Class I
Class II
RECORD CHANGE ONLY

MFG EFFECTIVITY:
(Fiscal Week, S/N or Lot No)

Reason For Change:

MISC ITEMS AFFECTED	YES	NO
PRODUCT SPEC		
FIELD IMPACTED		
MANUALS		
SAFETY (UL/CSA)		
SOFTWARE		
SPARE PARTS		
TOOLING		

PART DISPOSITION CODES
1 USE AS IS
2 REWORK
3 RESIDUAL, RETAIN FOR OTHER USES
4 NOT APPLICABLE
5 NO EFFECT, DOCUMENT CHANGE ONLY
6 SCRAP
7 SEE NOTE, PAGE ____

DWG SIZE	DOCUMENT No AFFECTED	TITLE	REV FROM	REV TO	REPLACED BY	Next Order	At Vendor	In Production	In Check Out	In Finish Goods	In Stock On Hand	In Stock Spares

DESCRIPTION OF CHANGE:

Originator Date	Engineering Date	Manufacturing Date	Date	CM Mgr Date

Figure 7.2
Recommended engineering change order (ECO) form.

be initiated and a new number assigned to that change order. In order not to repeat valid information already existing on the original change order, a copy of that change order may be attached to the revised change order. The only revision which is permitted to a change order is for a change in effectivity which we will discuss in detail under item 6.

2. Date approved. When the manager of the change control organization has given his or her approval to this change, the date of that approval should be indicated.
3. Product identification number. As shown in Figure 7.2, both the old and new product identification numbers are shown, if in fact they are changed as a result of this particular change order. First, the affected old product identification number(s) is shown and then the new product identification number(s) is shown. This change to the product identification number must be approved by the sales or marketing organization of your company as they should be controlling these numbers. It is also possible that you will have more than one product identification number shown as more than one product can possibly be affected by this change. In order for this change order to be complete, all products affected by this change must be shown on the change order. Separate change orders for each product is not necessary as a given engineering change order should be a complete package in itself.
4. Class of change. In the form shown in Figure 7.2, three different classes of change are indicated. These classes are shown as a means of indicating whether a change is non-interchangeable, interchangeable or only a record change. A Class I change is considered a non-interchangeable change, a Class II change is an interchangeable type change, and a record change is only as the title indicates, a change to handle typographical errors on your document. See Chapter 8 for more information on interchangeability.
5. Miscellaneous items. The engineering change order form should contain a list of items which can be associated with your kind of business. These items become "ticklers" to insure that we get the *total* impact of any change. Some of the items shown in Figure 7.2 may be appropriate for your company and others may not. In any case, you should have a list of items which are appropriate for your company. For instance, your company may be a medical

device company and you are subject to adhering to Food and Drug Administration regulations within the federal government. An item in your change order form under miscellaneous items would then be "FDA affected." Remember, we are always trying to attain the total impact of a change.

6. Manufacturing effectivity. Manufacturing effectivity means that we are trying to determine *exactly* when a change was implemented in a product on the manufacturing assembly line. There are three possible ways in which we can do this: either by fiscal week, product serial number, or lot number. Fiscal week makes use of what is called a fiscal week calendar. An example is shown in Figure 7.3. The purpose of this type of calendar is to condense the normal Gregorian calendar such that every week on the calendar shows a complete seven days. When we use the fiscal week calendar, a week is shown as a four digit all numeric number, such as 9515. The first two digits such as 95, indicate the last two digits of the year that the change is effective, such as 1995. The last two digits of 9515 indicates the number of the week in the fiscal week calendar, such as 15, that the change is projected to be installed in. So the number which would go in the effectivity block of the change order form would be 9515. If it is possible to insert the first serial number of the unit in which the change is to be installed, that number would be used in the effectivity block of the change order form.

   But sometimes it is impossible to serialize certain kinds of very small products, if that is the case we can usually identify a group of these small products by what is known as a lot number, assigned by your production control organization. This again is no more than an all-numeric type number, such as lot number 423. That would mean a certain number of your small products have been grouped together as a lot and are now identified for change and manufacturing control purposes as lot number 423. That number would be inserted in the effectivity block on the change order.

   If a particular change impacts more than one product, you may have to show several different effectivities for each type of

First Quarter

Mo	Wk. No	Period	MON	TUE	WED	THU	FRI	SAT	SUN
J	1	I	30	31	1	2	3	4	5
A	2	II	6	7	8	9	10	11	12
N	3	III	13	14	15	16	17	18	19
1st Mo.	4	IV	20	21	22	23	24	25	26
F	5	I	27	28	29	30	31	1	2
E	6	II	3	4	5	6	7	8	9
B	7	III	10	11	12	13	14	15	16
2nd Mo.	8	IV	17	18	19	20	21	22	23
M	9	I	24	25	26	27	28	29	1
A	10	II	2	3	4	5	6	7	8
R	11	III	9	10	11	12	13	14	15
3rd Mo.	12	IV	16	17	18	19	20	21	22
	13	V	23	24	25	26	27	28	29

Second Quarter

Mo.	Wk. No.	Period	MON	TUE	WED	THU	FRI	SAT	SUN
A	14	I	30	31	1	2	3	4	5
P	15	II	6	7	8	9	10	11	12
R	16	III	13	14	15	16	17	18	19
4th Mo.	17	IV	20	21	22	23	24	25	26
M	18	I	27	28	29	30	1	2	3
A	19	II	4	5	6	7	8	9	10
Y	20	III	11	12	13	14	15	16	17
5th Mo.	21	IV	18	19	20	21	22	23	24
J	22	I	25	26	27	28	29	30	31
U	23	II	1	2	3	4	5	6	7
N	24	III	8	9	10	11	12	13	14
6th Mo.	25	IV	15	16	17	18	19	20	21
	26	V	22	23	24	25	26	27	28

Third Quarter

Mo	Wk. No	Period	MON	TUE	WED	THU	FRI	SAT	SUN
J	27	I	29	30	1	2	3	4	5
U	28	II	6	7	8	9	10	11	12
L	29	III	13	14	15	16	17	18	19
7th Mo.	30	IV	20	21	22	23	24	25	26
A	31	I	27	28	29	30	31	1	2
U	32	II	3	4	5	6	7	8	9
G	33	III	10	11	12	13	14	15	16
8th Mo.	34	IV	17	18	19	20	21	22	23
S	35	I	24	25	26	27	28	29	30
E	36	II	31	1	2	3	4	5	6
P	37	III	7	8	9	10	11	12	13
9th Mo.	38	IV	14	15	16	17	18	19	20
	39	V	21	22	23	24	25	26	27

Fourth Quarter

Mo.	Wk. No.	Period	MON	TUE	WED	THU	FRI	SAT	SUN
O	40	I	28	29	30	1	2	3	4
C	41	II	5	6	7	8	9	10	11
T	42	III	12	13	14	15	16	17	18
10th Mo.	43	IV	19	20	21	22	23	24	25
N	44	I	26	27	28	29	30	31	1
O	45	II	2	3	4	5	6	7	8
V	46	III	9	10	11	12	13	14	15
11th Mo.	47	IV	16	17	18	19	20	21	22
D	48	I	23	24	25	26	27	28	29
E	49	II	30	1	2	3	4	5	6
C	50	III	7	8	9	10	11	12	13
12th Mo.	51	IV	14	15	16	17	18	19	20
	52	V	21	22	23	24	25	26	27

Figure 7.3

Example of a fiscal week calendar. Source: AT&T.

product indicated in the product identification number block. This is why a fairly large space is left for the manufacturing effectivity on the change form in Figure 7.2. The important thing to remember about effectivity is that we must know exactly which product, by serial number or lot number, had this change implemented in it. For effectivity purposes, you will eventually have to change the fiscal week number downstream to either a specific serial number of a product or a lot number and hence, the fiscal week number really only becomes a temporary effectivity for processing this change.

7. Reason for change. This is probably the most important part of your engineering change order form because it is here that we will describe why we need this change, which is a very important part of determining the total impact of a change. Simple statements such as manufacturing cost reduction or maintenance cost reduction or other simplified statements really do not depict the total reason for a change. You can ask yourself the question; what items, what parts, what functions are being changed in order to reduce the cost? What happened? Where is the problem? These are the questions that ought to be asked and answered in this area. A complete statement of the reason for change should be inserted in this area so that everyone can understand why the change is needed, what the problem is, and why we are going to the trouble and expense of generating an engineering change order to handle the processing of this proposed change.
8. Documents affected. As shown in Figure 7.2, all documents affected by this change should be indicated in this area, by part number and description. If it is a non-interchangeable change (Class I), then the new part number ought to be indicated under the column headed by the words "replaced by." If it is an interchangeable change (Class II), then the current and new revision letter status should be shown. All documents affected by the change must be listed, including the bill of material if impacted by this particular change.
9. Part disposition. Part disposition is one of the most important parts of an engineering change order form as we must know what people are suppose to do with the parts on order, parts at the vendor, parts in production, parts in inventory, and so on, wherever

the parts are located. Whether the headings shown under the part disposition area of Figure 7.2 are appropriate for your kind of business is for you to determine. Possibly other terms must be used; the important thing is that we must know how to dispose of all the parts affected by this change wherever they may be located. In another area of the form is a list of codes which can be used in the part disposition area in order to condense our form. Again, these codes may be appropriate or inappropriate for you and your kind of business. But this is one way to indicate what has to be done with the parts impacted by this change. It is amazing how many companies do not even include parts disposition on their engineering change order form. They then wonder why they have such a huge parts inventory, some of which is obsolete.

10. Description of change. In this area we must describe what has to be done to the documents affected by this change. A lot of companies will use a from-to type of description to indicate what has to be done. It is much preferred that actual copies of the documents affected by the change be attached to the change order form and the documents actually *marked-up* to indicate exactly what has to be changed. For example, a dimensional change must go from 1/4 to 1/2 inch or the material must change from steel to aluminum or whatever the change might be. Marked-up documents are much preferred as this eliminates a lot of potential errors. It should also be noted that we do not have to make a copy of the entire document and attach it to the change order as large size drawings may make the change order package much thicker than necessary. With the use of some type of copying equipment, it is possible to make a copy or reduced copy only of the area affected on a document and hence, we may be able to use a normal 8 1/2 x 11 sheet of paper attached to our change order package. It is also possible if we are not recording the title block on that marked up document that we may have to note the document number on the marked-up document for tracking purposes.
11. Approval of change. As shown in Figure 7.2, only four signatures are really required for approving a change. Namely, the originator of the change, the engineer responsible for the design of the product, and the manufacturing engineer responsible for the producibility of the product. Last but not least, is the signature of the

manager of the organization, responsible for processing changes. His or her signature on the engineering change order would indicate that *all* activity regarding a particular change order has been accomplished. We have discussed the use of only four signatures on the engineering change order form. This is the recommended number of signatures. If your company has less than this number, that is good. The fewer the approval signatures on your forms or documents the faster the processing time for approval of documents and changes. More than four approval signatures on a change is in most cases unnecessary. People should step back and take a careful look at the necessity for having a large number of approval signatures. The author has heard of some companies which have ten or eleven signatures on their documents and change forms. It is entirely possible that those signatures are absolutely necessary. On the other hand, signatures usually indicate the organization's approval of a document or change. And is it possible that some of these signatures can be combined under one signature, such as having the person responsible for the produceability of the product responsible for representing the quality organization, the production organization, the production control organization, the purchasing organization, and so on?

7.3 Configuration or Change Control Board (CCB)

In order to reduce the number of signatures, consider making use of what is called a configuration or change control board (CCB). Some companies which use a CCB have found it beneficial but only if they have a chairman who has been trained properly or is experienced in leading a board or a committee, as this group might be called. If a CCB is used, it can work very effectively in the overall processing and approval of a proposed change. But only if managed properly. The person responsible in your company for processing changes is usually either the chairman of the CCB or the secretary. In any case, it is highly recommended that a knowledgeable and forceful individual be the chair of the CCB in order for it to work effectively. That person can be given the authority to approve the change on behalf of the CCB. The CCB can be a very effective group if it meets on a regular basis, such as daily, twice a week, weekly, or monthly. On the other hand, it is possible that the CCB may be composed of the members of the product design team, see Chapter 2. If that group is used as the change control board, you have the best representation for such a board as it should be composed of representatives of all the major functional areas of your company. This group is responsible for reviewing and possibly approving all non-interchangeable changes which occur in your company. It is not necessary for them to review interchangeable or record change only type changes.

7.4 Engineering Change Cost Estimate

Along with completing the engineering change order form, it is also extremely important that we gather the necessary costs associated, especially with a Class I non-interchangeable change. Figure 7.4 is an example of what is called an engineering change cost estimate form. Through the use of this very complex form, it is possible to accumulate the *total cost* of a proposed non-interchangeable change. It is possible that for your company, there are some items shown which are not appropriate. On the other hand, you might have additional items which are appropriate and hence, this form should be modified accordingly. But this is an attempt, by using this form, to *accumulate all costs* associated with a proposed non-interchangeable change. Now, we do not have to accumulate costs of interchangeable changes or record type changes as these costs are usually considerably more or less standard, than a non-interchangeable change. Typically, a non-interchangeable change can be a very high cost to a company and hence, we must know those costs. The process for accumulating engineering change costs should be done in parallel with the development and review cycle for a proposed change so as not to delay the overall change processing system. The gathering and approval of change costs must be completed before release of approved changes to manufacturing.

Once all of the costs have been accumulated and the total cost determined for the change, approval of those costs must be obtained from company management. It is suggested that a company policy exist not only on your change control system but also on who can approve the total cost of a change.

ENGINEERING CHANGE

-COST ESTIMATE-

-EXAMPLE-

PRODUCT	DATE	ECO

	MAN HOURS	RATE OR PART COST	LABOR/MAT'L O H %	COST	ESTIMATED BY
A. CHANGE ORDER IMPLEMENTATION					
1 DRAFTING DRAWING CHANGE					
2 MANUFACTURING PROCESS CHANGE					
3 QUALITY ASSURANCE PROCESS CHANGE					
4 MANUAL (S) CHANGE					
5 TOOLING CHANGE					
6 SOFTWARE					
8 TOTAL					
B. RETROFIT COSTS					
1 FIELD UNITS					
A PARTS COST/UNIT					
B MFG LABOR COST/UNIT					
C CUSTOMER SERVICE COST/UNIT					
D. ENGINEERING COST/UNIT					
E. TOTAL COST/UNIT					
F. NUMBER OF UNITS					
G. TOTAL (MULTIPLY ITEM B1E TIMES B1F)					
2. FACTORY UNITS					
A. PARTS COST/UNIT					
B. MFG. LABOR/UNIT					
C. TOTAL COST/UNIT					
D. NUMBER OF UNITS					
E. TOTAL (MULTIPLY ITEM B2C TIMES B2D)					
C. INVENTORY SCRAP AND TOOLING REPLACEMENT					
1. TOOLING					
2. COMPONENT PARTS					
3. MANUFACTURING PARTS					
4. MANUALS					
5. SPARES					
6. TOTAL CHANGE ORDER COST					

D. **TOTAL CHANGE ORDER COST** (ADD ITEMS A8, B1G, B2E & C6. ENTER TOTAL HERE)

E. **PRODUCT MFG. COST CHANGE**	PRESENT	FUTURE	CHANGE (+ OR -)	EST. BY
1. PARTS COST CHANGE/UNIT				
2 MFG. LABOR COST CHANGE/UNIT				
3. TOTAL COST CHANGE/UNIT				
4. NO. OF UNITS NEXT 12 MO'S.				
5. TOTAL PRODUCT COST CHANGE (MULTIPLY ITEM E3 TIMES E4)				

INSTRUCTIONS: ITEMS OF INFORMATION TO BE ENTERED BY GROUPS INDICATED BELOW.

APPLICATIONS OR PROGRAMMING OR SOFTWARE - A6 DRAFTING - A1
PRODUCT IMPROVEMENT OR CONTINUATION ENGINEERING - B1D
MANUFACTURING OR PRODUCTION - A2, A5, B1B, B2B, C1, E2,
MATERIALS OR MATERIAL CONTROL OR PRODUCTION CONTROL - B1A, B2D, C2, C3, E1, E4, B2A
PUBLICATIONS - A4, C4 CUSTOMER ENGINEERING - B1C, B1F, C5 QUALITY ASSURANCE - A3

APPROVAL DATE

APPROVAL DATE

CHANGE BOARD CHAIRMAN DATE

Figure 7.4

Example of an engineering change cost estimate form.

An example of what that dollar value is for various levels of management is shown below:

The manager of the change processing system should be capable of approving changes up to a figure of five thousand dollars. His manager should then be able to approve change costs accruing to ten thousand dollars. The executive of engineering should then be capable of approving change costs accruing to twenty five thousand dollars or more. And last but not least, the chief executive officer or executive vice president of your company should be able to approve all change costs which total more than fifty thousand dollars and anything over one hundred thousand dollars would require approval of the president of your company. Remember these are only ballpark figures, possibly for a company of less than five hundred employees. Over five hundred employees or more, the dollar figures could be much greater.

7.5 Deviations/Waivers

In any good overall engineering change system it has been found by most companies that some kind of *deviation* and *waiver* process is usually required. So what do we mean by these two terms?

A *deviation* is a short term or temporary departure from an approved engineering standard, specification, drawing or other engineering document of a company designed and manufactured or purchased product and is effective on one or more of its serial numbers.

Typically, a *deviation* is used in an emergency situation only. It is for a given number of units and for a given period of time. It is not -- repeat -- it is not an authorization to change any engineering documents. But it can become a request for engineering change such as the ECR form previously discussed. It is not an authorization to change any products which are out in the field at a customer site. And last but not least, no revisions can be made to a deviation. If a mistake is made, a new deviation must be generated. Figure 7.5 is an example of a deviation form which might be used by your company.

The important things to include on a deviation form are:

1. A deviation number which can be used for tracking purposes.
2. The product identification number and effectivity of the deviation.
3. Listing of the serial numbers affected by this deviation.

LOGO	DEVIATION AUTHORIZATION		DEVIATION NO D
	AFFECTED PRODUCT NO	EFFECTIVITY (FISCAL WEEK AND/OR S/N)	SHEET OF
		QUANTITY AFFECTED	DATE
		PROJECTED CLEARING DATE	CLASS ☐ I ☐ II

DEVIATION AFFECTS			YES	NO	ECO / FCO NUMBER
☐ MANUALS	☐ EQUIP FUNCTION	ECO REQUIRED	☐	☐	________
☐ SOFTWARE	☐ MAINTAINABILITY	FCO REQUIRED	☐	☐	________
☐ DIAGNOSTICS	☐ SPARE PARTS				

REASON FOR DEVIATION

PARTS/DOCUMENTS AFFECTED (INCLUDING MANUALS)

PART//DOCUMENT NUMBER	DESCRIPTION	DOCUMENTS ATTACHED	
		YES	NO

DESCRIPTION OF DEVIATION

D

APPROVAL

ORIGINATOR DEPT DATE	ENGINEERING DATE	MANUFACTURING DATE

Figure 7.5
Example of a deviation form.

4. Indicate whether or not an ECO might be required.
5. Indicate the class of deviation. If the deviation impacts the documentation or configuration, it is a Class I deviation. If it does not impact the configuration, it is a Class II deviation.
6. The reason for the deviation should be defined in a clear statement.
7. A listing of the parts or documents affected by this deviation should be shown.
8. The approval of this document should be the same as for an engineering change.

A *waiver* is different from a deviation as shown in the following definition:

A *waiver* is a long term or permanent exemption from the requirements of a product design standard, a procedural or process standard, specification, drawing, or any other engineering document used in a company designed and manufactured or purchased product.

Again, a *waiver* is not an authorization to change any engineering documents and it is not an authorization to change any products in the field at a customer's site. Typically, a *waiver* is described in the overall product design objectives type document. Once the product design objectives are approved by management, that is your approval for a waiver and no special waiver form is needed.

7.6 Summary

We have discussed the forms required for a good engineering change control system including an engineering change request, engineering change order, engineering change cost estimate, and deviations and waivers. It should be noted that an approved ECO, when complete, really becomes an ECO package which includes the ECO form and all affected marked-up documents. The ECO package should be so complete that engineering can revise all the affected documentation; that manufacturing can revise their processes and install the change on the assembly line without any approved and released revised documentation; and that other company functions can perform their change responsibilities as defined in the ECO package or in the change processing system.

All forms used in your change processing system should be *simple, easy to understand, and user friendly.* Whenever changes are contemplated to these forms, it is imperative that the users of these forms are adequately represented during the initiation and review of the revised forms *before* approval.

One last thing on forms. The author has seen many, many different ECR and ECO forms. The most glaring omission on these forms is parts disposition, which probably is one of the most costly parts of an ECO package. On the other hand, a lot of companies have priority classes shown on an ECO form which in the author's opinion is unnecessary. The classes of interchangeability as described in Chapter 8 really give you all the priority you need. All Class I non-interchangeable changes are always top priority. Class II changes are next and record change only changes come last.

Chapter 8
Interchangeability

In teaching this topic in my seminars and also through various consulting assignments, my experience has shown that many companies, especially "hi-tech" companies, have not thought about interchangeability rules. It is my conviction that a good engineering documentation control system must contain a set of interchangeability definitions and rules if we are to control our engineering documentation. Hence, this chapter will show the absolute necessity for having such a set of rules.

8.1 Class I Change

Let us now define the main categories or classes of interchangeability:

Non-interchangeability (Class I change) occurs when a part is to be changed and the changed part will not physically fit or function as a replacement for the unchanged part in *all* applications of its previous revisions, and all unchanged parts will not physically fit or function as a replacement for the changed part.

A *non-interchangeable change* (NIC) requires that new part numbers be assigned to the affected parts or assemblies which are not interchangeable.

This complete definition, though complex, is one of the hardest items for people to accept when they have a non-interchangeability parts situation. It is an absolute requirement that we cannot go "around left end" to avoid assigning new part numbers when a non-interchangeable situation occurs. Using letter revisions for a part number to track non-interchangeability will never work, except for only one situation. That situation occurs when it is possible to control all parts and assemblies in-house affected by the non-interchangeable change. In other words, those affected parts and assemblies are all located in-house and no product has been shipped out the door and no parts exist in our spare parts inventory. As long as we have complete control of all parts and assemblies affected by the change, we can update the revision letter of the part number, e.g., from A to B or B to C without changing the part number. This is the *only* exception to the rule for changing part numbers for a non-interchangeable change. Now here are a few examples of when part numbers may or may not be changed:

1. If you have product in the field that must change because of a non-interchangeable change, part numbers will have to change such that we can track the product configuration in the field.
2. Assemblies not field replaceable (spared) need not change part number on a non-interchangeable change.
3. The top level assembly need not have its part number changed for a non-interchangeable change providing your overall documentation or configuration system uses some type of a mod system to the product identification number to track major non-interchangeable changes to your product, such as 6632-A01, wherein A changes to B for external interface non-interchangeable changes and 01 changes to 02 for internal interface non-interchangeable changes.
4. "Form type" and "safety type" non-interchangeable changes should not be allowed unless the person responsible for the prod-

uct design is willing to revise the product specification to add specific criteria related to these items.

5. As stated previously, part numbers need not change on an non-interchangeable change if all the parts are under your company's control in-house and if no parts have been produced and shipped.

8.2 Class II Change

Now let us look at the definition for *interchangeability*:

Interchangeability (Class II change) occurs when two or more parts possess such functional and physical characteristics as to be equivalent in performance and durability and are capable of being exchanged without alterating the parts themselves or adjoining parts except for adjustment, and without selection for fit or performance.

If a given situation occurs and the parts affected fulfill this definition then it is only necessary to upgrade the revision letter of the part and document number for the affected documents or parts.

8.3 Record Change Only

A third class of change is called *record change only* and its definition is as follows:

A *record change only* is made to correct non-configuration affecting errors on the document or to add a tab to an engineering document. It must not affect fit, form or function.

A *record change only* requires a revision letter update to the document and there is no change to the part number.

It should be noted here that anytime we alter an engineering document in any manner, that as a minimum, the revision letter must be updated, for example, from A to B, B to C, etc.

8.4 Reliability Interchangeability

A special type of an non-interchangeability situation is called *reliability non-interchangeability*. It is defined as follows:

Reliability non-interchangeability occurs when a part or assembly is changed to improve the reliability to the level necessary to meet the intent of the product specification.

An example of this type of non-interchangeability would be the use of a bigger and better light bulb in your product. For instance, if you have a light bulb which has an expected life of only 25 hours it will keep the mean time between failure (MTBF) of your product very high. Subsequently, if you should find a light bulb which has an expected life of 1000 hours, this would be a very considerable improvement. The MTBF for your product would then increase substantially with the use of the 1000 hour light bulb. But, from a interchangeability standpoint, the bigger and better light bulb of 1000 hours of life would receive a different part number from the 25 hour light bulb, even though they might fit and essentially function in the same way, due to the reliability factor. Remember, if we have a tabulated drawing for the light bulb, changing the tab part of the part number is the same as changing the total part number.

Another situation which deserves special mention on the subject of interchangeability is what might be called "the printed circuit card assembly syndrome." This syndrome says that any change to a printed circuit card assembly is a functional change and therefore it must be non-interchangeable and we must change part numbers. This statement is absolutely *wrong*! The real question is: Does the problem to the printed circuit card assembly create a functional change to meet the product specification or to improve the specification requirements above what was originally intended? The point is that we do not have to change part numbers necessarily for *all* printed circuit card assembly changes.

8.5 Differentiating Between Interchangeable and Non-Interchangeable Changes

In order to assist you in determining whether you have an interchangeable or non-interchangeable situation with a given problem, a truth chart is shown in Figure 8.1. In this figure we ask several questions which really are only the definitions that we have already given you for interchangeability and non-interchangeability. The first

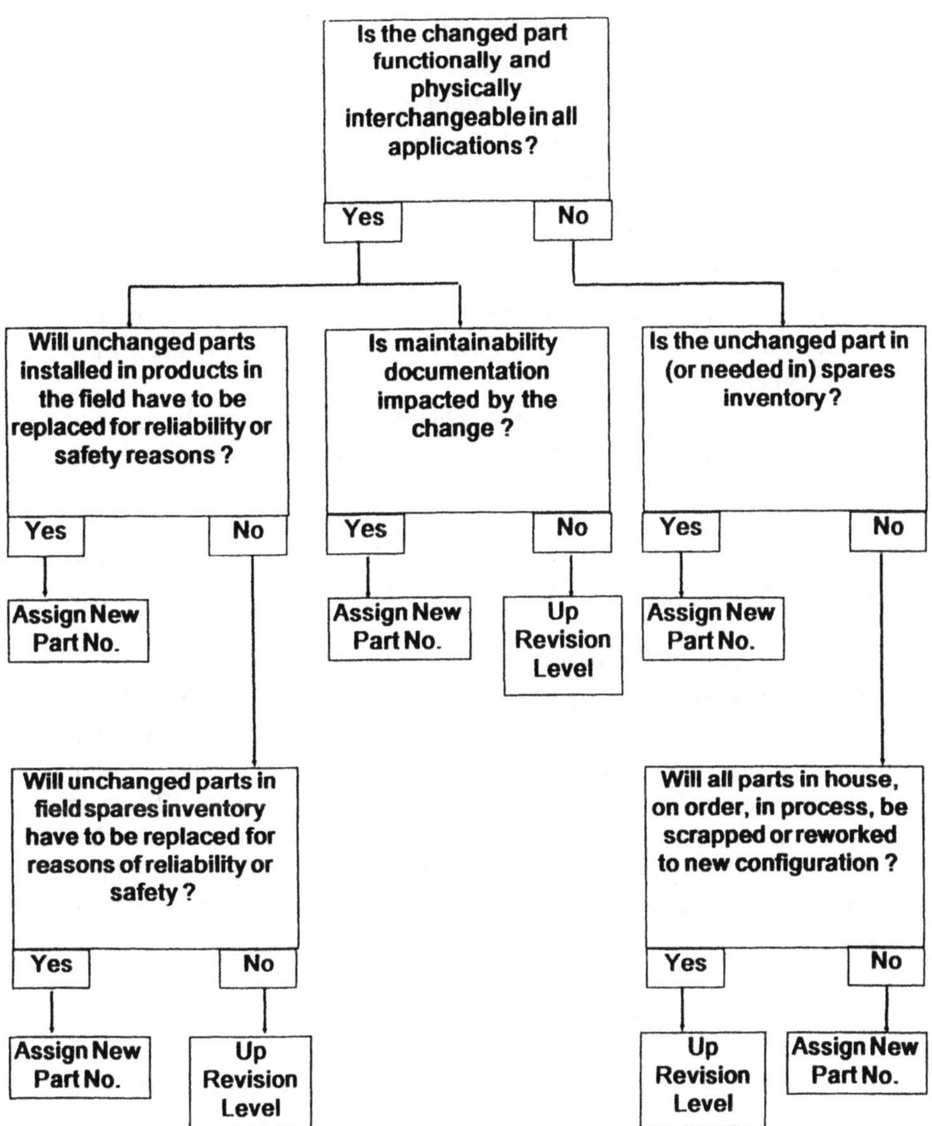

Figure 8.1

Parts interchangeability analysis chart. Whenever a new part number is indicated, the next higher assembly level must be similarly analyzed to determine if it is a part number or revision level change.

question asked as shown in Figure 8.1 is: Does the change part work functionally and physically interchangeably in all applications? If it does we ask ourselves some additional questions and eventually we can determine whether or not we must assign a new part number for a non-interchangeability situation or upgrade the revision level status of the part number, which would mean we have an interchangeable situation. Remember one thing in analyzing a given problem situation, whenever a new part number is required because of non-interchangeability being affected, we must always review the next higher assembly levels in the family tree or bill of material listing for a product in order to determine whether or not a new part number must be assigned.

It is important in this analysis that we look at all applications of the affected part number or part numbers. Hence, it is imperative in our bill of material system that we have a *where-used file* for all parts and assemblies listed in the bill of material. If a part or assembly is non-interchangeable in only one application, the part number must change but you can retain the old part number for all other applications.

In making your analysis of a given problem situation, wherein you are unsure of having a interchangeable or non-interchangeable situation, it is possible that you could ask yourself one more question that may assist you in defining this particular situation. That question is: *Do you care*? If you care that product A, serial number 122, as an example, does *not* have this proposed change in it but serial number 510 of this same product *does* have the proposed change installed and you care that you need to know the difference, then you must change part numbers to reflect that change in the product configuration.

8.6 Interchangeability Rules

Appendix A1 includes a complete set of interchangeability rules which your company may want to use. You will notice that terms are defined again in these rules along with all the other possible rules that you may need in trying to analyze a given interchangeability or non-interchangeability situation.

8.7 Serviceability or Maintainability

Another item which must be considered when reviewing an interchangeability situation is a term called *service level*. It is defined as follows:

The *service level* of a product (serviceability or maintainability) is that level on the branch of the family tree or bill of material of a product where it is most economical or practical to spare or service a part or assembly.

If you refer to Figure 4.4 under configuration identification, you will see the structure of a generic type family tree for a product and its documentation. As you move down the family tree from the top, you start with the top level assembly of the product and going down to the next major sub-assembly level you may have several major sub-assemblies which go into making up your particular product. This is the first level on your family tree. As you go down the structure of the family tree of your product you may have additional sub-assemblies which make up these major sub-assemblies, and not only additional sub-assemblies but parts also. You must determine for your particular product how it is going to be serviced such that we can determine which items must be spared for our field operation and subsequent spare parts inventory. So what is meant by a spare part?

8.8 Spare Parts

A *spare part* is defined as an individual item (such as piece part or assembly) which is normally stocked as a replacement for a failed or worn out item of a delivered product to a customer.

By analyzing the family tree or bill of material structure of a product, we can determine which items should be spared within your product.

8.9 Spare Parts Provisioning

Spare parts provisioning is the process of determining the types, quantity, and stock point locations of parts to be spared and required to support and maintain a product when in service or use.

An important activity related to defining which items are to be spared for your product is called *provisioning*. Determining who is responsible for spared items for your product is important. Spare parts provisioning is a *joint* responsibility of the design engineering organization along with the people responsible for your products in the field or at a customer site. Sometimes this group is referred to as your field engineering or customer engineering group. In any case, they are the most knowledgeable about what is occurring to your products in the field or at a customer site. The provisioning activity is an extremely important one and both of these organizations must determine which items in your product are to be provisioned as they must include whether the items are considered repairable or non-repairable at a customer site or at a regional repair depot, or possibly at your factory. Whether a part or assembly is determined to be a spare part is a very crucial factor in determining whether or not a problem to a product which requires a change order to be processed will be considered interchangeable or non-interchangeable.

8.10 Summary

Determining the interchangeability status of the parts and assemblies involved in a given product problem situation, which requires an engineering change order to be processed, is probably one of the key factors in controlling your engineering documentation. If the situation is such that you must change part numbers because of a non-interchangeability situation, it is imperative that you change those part numbers and not try to go "around left end" and update the revision letter status of the part number. Usually when we are updating revision letter status for a non-interchangeable situation we are doing so because of costs. We all know that changing part numbers can be very expensive. On the other hand, not changing part numbers can be even more expensive. It really gets down to " pay me now or pay me later." I can assure you, that if you do not change part numbers for a non-interchangeable situation, you are asking for more trouble than you can possibly expect. Though cost is an extremely important factor, when we make mistakes somehow we must pay for them. There is only one way to control the status of our documentation for a product and that is by changing part numbers for non-interchangeable situations. It is in fact the only *absolute* way of controlling our documentation.

Chapter 9
Analyzing a Proposed Engineering Change

So far, we have discussed various aspects of an engineering change, from the philosophy of a change and the need to get the total impact of a change to the various forms that might be used in an engineering change system, including interchangeability. Now, let us discuss a technique for analyzing a proposed engineering change. Not every engineering change must be subjected to the discussion of the material in this chapter, primarily only non-interchangeable changes.

When we start to analyze a proposed non-interchangeable change to a product there are many, many factors to determine. First, do we need the proposed change? Do we have the right solution and are we trying to accomplish the right thing as far as this proposed changed? Let us now look at Figure 9.1, which is a truth chart of the various possible steps that a change may be subjected to. As we review Figure 9.1, please note that the first question asked is: Does the product's configuration or documentation *meet* the product specification? In other words, does this product perform the way it was intended to perform when we originally conceived the product? Also, is the product

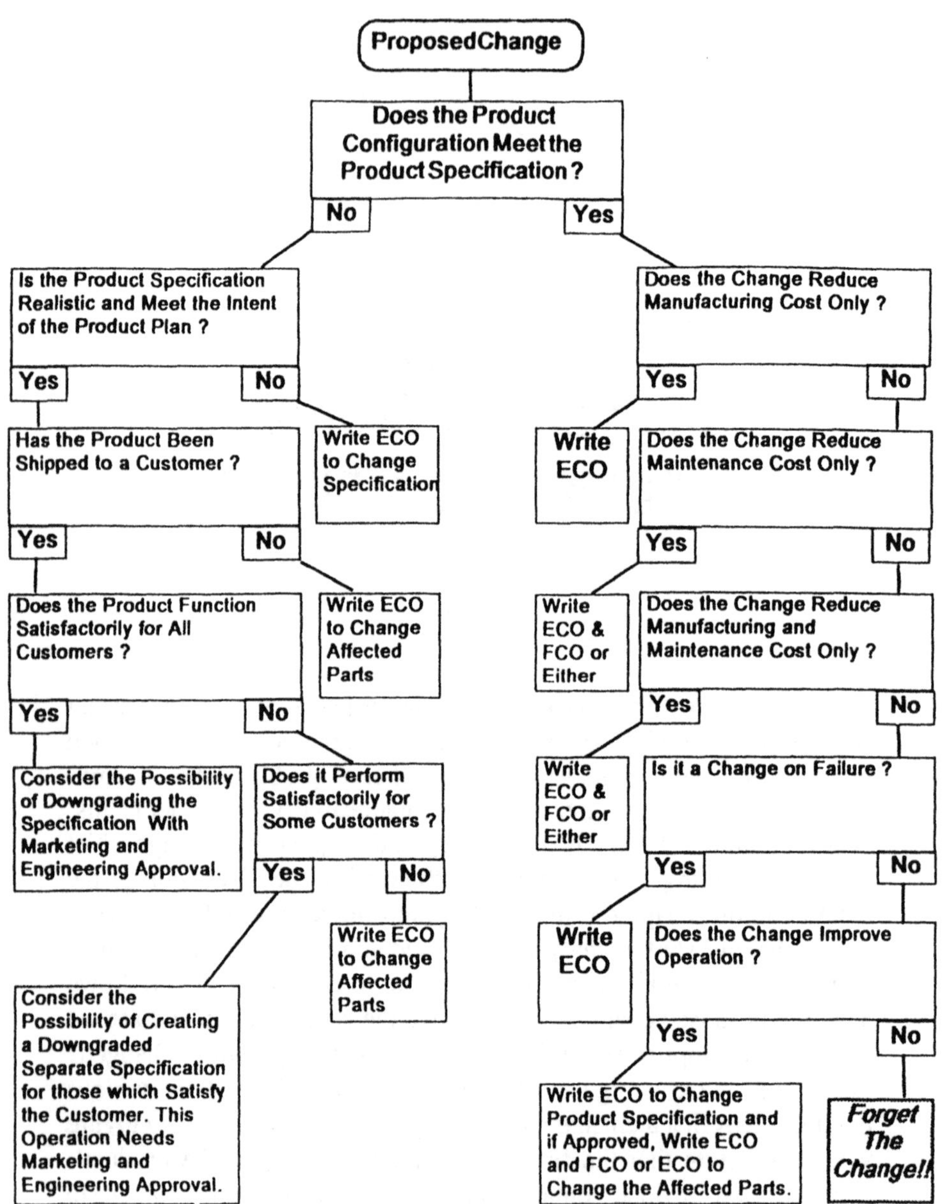

Figure 9.1
Analysis of a proposed change to a product.

performing in the field at a customer site the way the customer expected when he purchased this product? If the answer is *no*, the product configuration does not meet the product specification, and then we should immediately ask ourselves is the product specification realistic and does it meet the intent of your product plan or product design objectives originally established in the conception stage of this product. If it does not, then we must write an ECO in order to change the product specification. If the product specification is realistic and meets the intent of your original product objectives, we have another question to ask. Has the product been shipped to a customer? If a product has been shipped to a customer, we must be very much concerned about customer satisfaction. If no product has been shipped to a customer, again, we can write an ECO to change the parts affected by this particular proposed change.

The next question to be asked is does the product function satisfactorily for all customers? This is an excellent question because you may be using a given appliance in your home only partially, while other people may be using it to its full capability and still others who are trying to use it for more than what was intended. So if the product is performing satisfactorily for all customers we can move on and say we might want to consider downgrading the specification with marketing and engineering approval. We would do this because the product is functioning satisfactorily for all of our customers and there is no reason to make any changes to our product. Hence, we may have established some requirements in our original product specification which are difficult to meet and hence, we would want to downgrade those requirements per our experience with the product in the field. If the product is only performing satisfactorily for a few customers, we may want to consider the possibility of creating a downgraded separate product specification just for those customers and possibly creating an ECO to change any parts that need to be corrected such that the balance of the customers can have a product which is operating satisfactorily.

Now look back at Figure 9.1 and the original question of does the product configuration or documentation meet the product specification. If we say yes, then we ask ourselves a series of questions to determine whether to write an engineering change for this problem. The first question asks, does the change reduce manufacturing cost only? This means that our manufacturing cost is possibly high and we desire to cut our cost such that we have a more profitable product. If the change is to reduce manufacturing cost, we will have to write an engineering change order. On the other hand, if we say no, then we ask whether the change reduces maintenance cost only? Again if the answer is yes, we will have to write an ECO, and if you have product in the field you may find that you have to have some kind of a field change system, which we will discuss in Chapter 11. If the answer is no, it is not just a maintenance cost reduction, then we ask, Is it a change to reduce both manufacturing and maintenance cost? If the answer is yes again, we would write an ECO (engineering change order) and also possibly an FCO (field change order). If the answer is no, then it is not to reduce manufacturing or maintenance cost but it is a *change on failure*. In other words, when the product is operating in the field and we have a problem that must be repaired, but it can be repaired at the time the product fails, this would be called a change on failure. If it is a change on failure, again we would have to write an engineering change order. If the answer is no, and it is not a change on failure, then we ask the last question, Will this change improve the operation of our product? If the answer is yes, we can write an engineering change order to revise our product specification and also to change any affected parts. If now we have answered all these questions and we've said no, the last question is, Does the change improve the operation? If you say no, maybe it's time we *forget the change*. You see, it is always possible that we can forget a particular change if we give it the proper analysis *before* the fact rather than *after* the fact. A lot of companies complain about the number of changes they process in a week, a day or a month. One of the many reasons they have such a problem is because they have not taken the time to analyze the proposed change in the first place to see if they really have a problem. So don't forget it is always possible to - *forget the change*!!

9.1 Summary

We have covered many aspects of configuration management, all having to do with the processing and analyzing of a proposed change. We have defined what a change is, we have given some philosophy related to change that should help in the managing of change. We also have discussed various forms that should be a part of any good engineering change system, and last but not least, we have discussed the subject of interchangeability, which is a must for any good engineering change processing system. And now you have some guidelines for actually analyzing a proposed change *before* the fact such that you do not spin your wheels over an unnecessary change. And now lets go on to the actual processing of a change.

Chapter 10
Change Processing System Requirements and Procedure

Whether you have a manual system or an automated system for processing proposed changes to a product, the major requirements for the engineering change processing system are the same. They are as follows:

10.1 Major Requirements

1. Limit changes. Somehow within your change processing system you must try to limit the number of changes. For instance, by analyzing a change in the first place as to its need, its cost and the feasibility of the proposed solution of the problem we may find that the change itself is totally unnecessary. In my seminars I have told people that if each engineering department had a six foot high neon light sign that reads, "Is this change really needed?", we could possibly limit the number of changes in our system at any time.

2. Preserve the company interest. We must always be aware when processing changes of the cost that might be incurred when processing or implementing a change. Hence, it is imperative that a cost analysis be made for at least all non-interchangeable changes so that we know what our total costs will be in the processing and installation of this change.
3. Rigid management control. Those of us associated with overall processing of a proposed change must insure that the rules and procedure for processing changes are adhered to. Exerting the necessary discipline such that people do not go "around left end" is absolutely necessary for a good engineering change processing system. Discipline is a key and management must be aware of the necessity for employees to adhere to the change procedure as defined by your particular company.
4. Clear definitions. As I stated initially in this book, a glossary of terms is an absolute necessity so that we are all using the same language.
5. Simple form. We have already described a typical engineering change form, the engineering change order. This is a rather simple form and that's the way it should be. The more complex your form is the more mistakes you will have in the processing of the change and the more cumbersome will be your overall change processing system. An example of a complex form might be to show too many classes of interchangeability. One company I know had about 10 different classes of interchangeability, primarily because of the bill of material or MRP system used. This is totally unnecessary. We must keep things as simple as possible in our forms as well as in our overall change processing system.
6. Ensure a closed loop. This means that we must have feedback from the manufacturing area when the change is first implemented on the manufacturing assembly line. We must know which unit or lot (by serial or lot number) first received this change so that we will know in the future that an old product does not have the change and a new product does.
7. Minimum cost. Your change processing system should be operated at the lowest cost possible. By keeping it as simple as possible, we will accomplish this particular item. The processing of changes is always a headache to your executive management be-

cause of the unusual high cost which sometimes occurs within the change processing system. So the advice here is to keep your system as simple as possible.

8. Status of changes. Most people today have some kind of an automated system, perhaps tied to the MRP system, which records the status of changes. Some of the items which should be considered in your record system are as follows:

   A. ECO number.
   B. Date initiated.
   C. Date received from originator.
   D. Date the proposed change was distributed for review.
   E. Date review completed.
   F. Date change completely approved.

   These items are the minimum required in your log for recording the status of proposed changes. Other information can be added but may become extraneous. In any case, try to have only a single record keeping system and don't have your change records located in two or three different areas.

9. Speed. Whatever kind of a change processing system you have, it should ensure the processing of changes expeditiously but not by sacrificing the accuracy of change information. Otherwise, "pay me now or pay me later" once again applies.

10. Ensure accuracy. This refers to the accuracy of the engineering change order information as well as to the accuracy of the items required for processing this change. As stated previously, we want to do it right the first time whenever possible. But we are all human and we are subject to making errors. Attaching marked-up documents to the ECO package will usually reduce your errors. There is no magic formula for reducing errors. We have to try to keep our errors to a minimum not only in the change package itself, but within the change processing system. Figure 10.1 is an example of a typical change processing system. It is not necessarily the epitome of all change processing systems but it does contain some of the elements which you should be concerned about in your system.

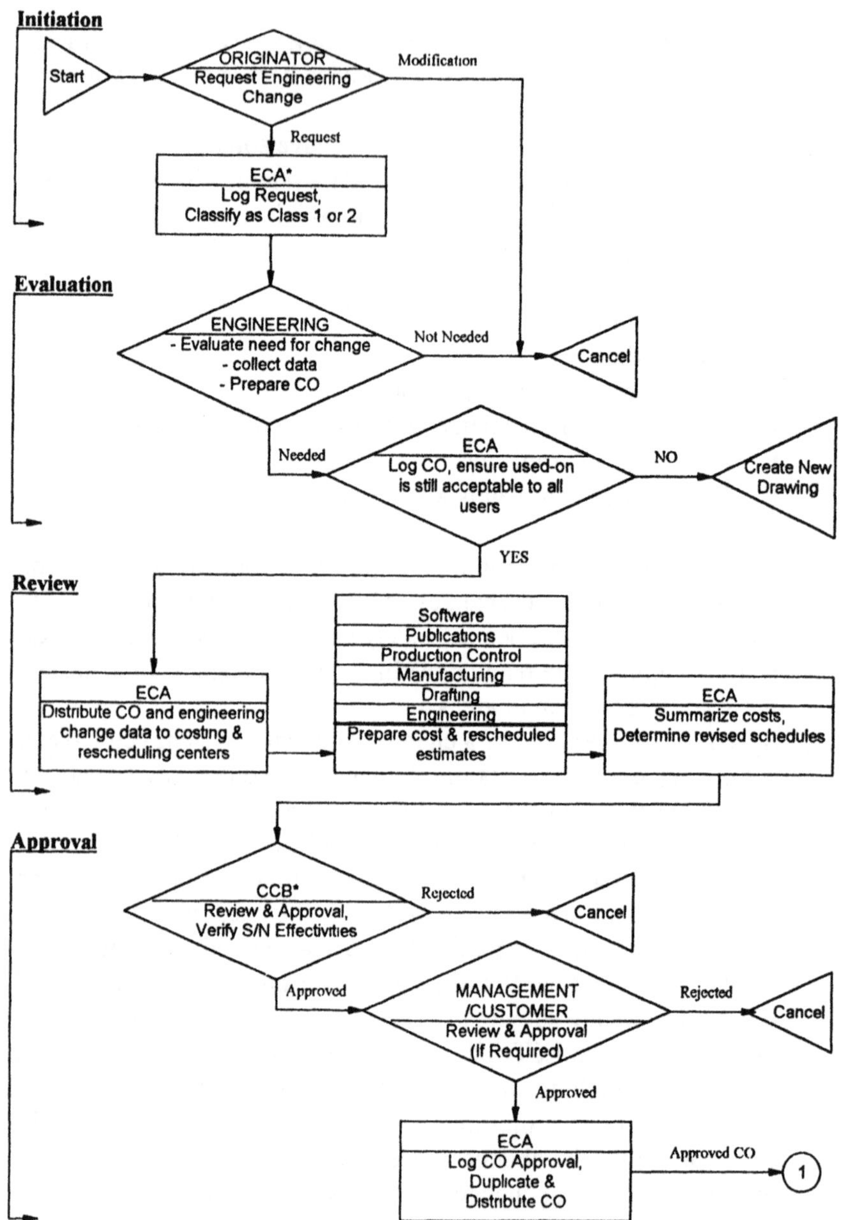

Figure 10.1A

Typical change processing system A: Implementation ECA: engineering change administrator, CCB: change control board, CO: change order.

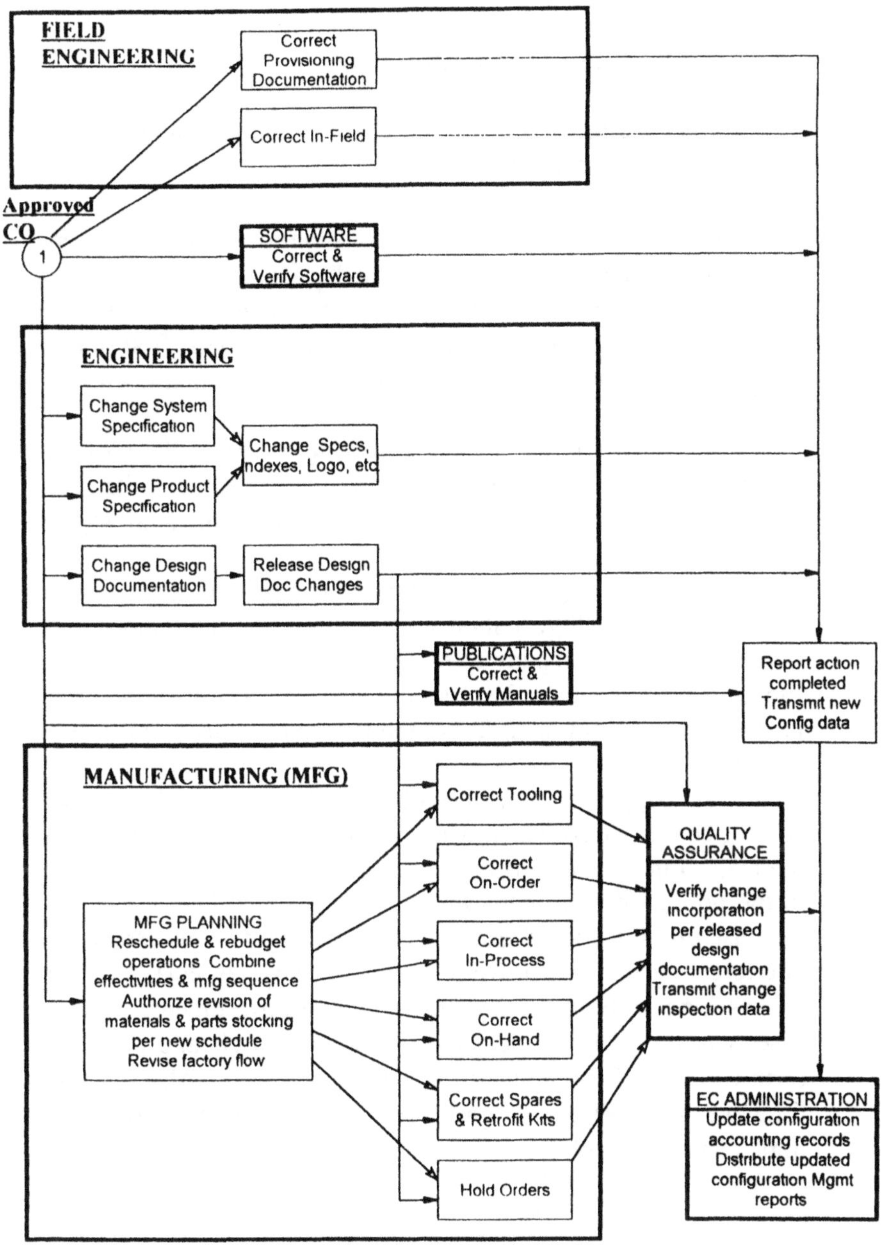

Figure 10.1B
Typical change processing system.

10.2 Major Change Procedure Ingredients

Any good change processing procedure should include the following five major ingredients:

1. Initiation. Somebody has to be responsible for requesting an engineering change in the first place, whether its the design or manufacturing engineer, an assembly person, or a manager, it makes no difference, but that's where your change starts. So the first step in your procedure must include initiation and where that document flows.
2. Evaluation. This is done by engineering to assess the specific problem which is discussed in the proposed change as to its need and best solution. Remember, we are trying to get the total impact of a change.
3. Review. Once the ECO form and package has been completed to the best of our ability, it should be reviewed by the major functions within your company as to its feasibility, its cost and how it can be scheduled into the manufacturing assembly line. The product design team or CCB are the best ones to accomplish this activity.
4. Approval. Approval is usually obtained by the use of a change control board (CCB) which usually consists of representatives in the major functional areas of your company. It can be equivalent to the product design team which was discussed in Section 2.3. Because of the nature of your business, it might be necessary for you to get customer approval of a given change especially if you are working with an agency of the federal government. In that case, you probably would have to complete a change proposal form called a ECP (engineering change proposal). Also, by this time we have hopefully gathered all the estimated costs for implementing this change, especially if it is a non-interchangeable change and we may have to get management approval of the change as to its cost. The ECP allows the government/ customer to also complete a change impact analysis before you, the supplier, change the item.
5. Implementation. Implementation consists of all the functions responsible for doing something as defined within the engineering

change order package. As an example, engineering is responsible for updating all the product documentation. Manufacturing is responsible for possibly correcting its tooling, correcting any manufacturing processes, correcting their spares and any retrofit kits which might be required and so on. The field or customer engineering group, which interfaces with customers in the field, is responsible for making any changes related to your customers. The publications department would be responsible for correcting and verifying the manuals or instruction sheets associated with your product. And last but not least, quality assurance will verify that the change has been incorporated per the approved engineering change order package or the released design documentation for this particular change. The group responsible for processing changes in your company would be updating all of their configuration accounting records and distributing copies of the change to the appropriate parties.

10.3 Tips for Streamlining a Company's Change Procedure

Listed below are some tips on how to streamline or "clean up" your existing change processing system:

1. Compare your system to Figure 10.1. If it is longer than this or more complex, try to determine how you can best streamline your system. Consider measuring the throughput time for a given change and then publish the results for review. Also, the use of statistical process control and quality management techniques will help you.
2. Consider streamlining your existing engineering change request and engineering change order forms as they may be too complex.
3. Consider developing or using an automated system for processing your changes.
4. Consider hand-carrying proposed critical ECOs for review and approval versus using the internal company mail system.
5. Insure change responsibilities for each major company functions are clearly defined.
6. Conduct a survey of your competition as to how they handle changes.

7. Are there too many functions involved in approving changes?
8. Obtain management awareness of those people who tend to go around "left end" of your change processing system. At the same time, try to determine why they are doing this, perhaps you have to change the system.

10.4 Summary

It is important to remember that processing proposed engineering changes from beginning to end is an extremely cumbersome process. Hence, the simpler we make our system the better. The less forms we have for processing changes and the less approval signatures required on a change the better. And last but not least, using marked-up documents will usually ensure the accuracy of your change such that we don't have to repeat ourselves. As stated previously, the speed that we use in processing a change reflects on the profitability of our companies. Hence, it is imperative that we keep our change processing system as simple as possible so that we can process our changes expeditiously.

Chapter 11
A Field Change System

Using a field change system is management's decision. It is entirely dependent on how you want to manage and control the maintenance aspect of your product in the field or customer site. There probably are many, many different ways in which we can handle changes to our products in the field, such as by having a formal field change system, either manual or automated; using some kind of field bulletin; use of a tech tip system or the use of a hot line fax system; or a combination of the above. If your company has made a decision on how to maintain your products in the field, this may mean that you have a group of field service engineers and you are prepared to correct problem situations directly at a customer site. On the other hand, you may desire to have repair of problem parts or assemblies done at designated regional repair centers. And last but not least, you may decide that all repair of problem parts or assemblies should be done at the factory. What type of field change system you need will depend on the type of maintenance effort of your product in the field .

Assuming that your company prefers a formal field change system, which includes maintaining your product at customer sites, then the following discussion would be appropriate. If your company prefers to do all of their maintenance at the factory or at designated regional repair centers, it may not be necessary to have a formal field change system and only field bulletins or hot line faxes are necessary.

11.1 Use of an FCO Form

In a formal field change system, a field change order (FCO), see Figure 11.1, can be used to describe the problem and solution. An FCO is defined as follows:

A *field change order* is the directive to install changes in a product after the normal manufacturing process in order that the product will perform to its written or implied specification, to meet safety standards, to improve reliability, or to reduce maintenance costs.

This definition does not include trying to improve the product or make it perform beyond its written or implied specification. It applies when, as an example, you have delivered a product to a customer and they expect it to do such and such and the product is not performing per its specification requirements. It will then have to be repaired using an FCO form in a formal field change system.

An example of the FCO form, which could be used in a formal field change system, is shown in Figure 11.1. Note that this form can be used for either what might be called a *regular* or a *selective field change*. A regular field change would be one in which all serial number units of the same configuration are impacted by the field change. A selective field change would be one in which only certain serial numbers of a given configuration are impacted by the change. A selective type field change is normally used only when certain customers need the change because of their application of the product and all customers are not required to have the change. A selective field change is a cost effective method of updating your product in the field if appropriate. In reviewing the FCO form in Figure 11.1, note that there is an identification number required on the FCO form. This number would be the same as the number of the ECO, which backs up this particular field change. The field change form also includes the following items:

1. The name of the product affected.

LOGO	CHOOSE ONE: ☐ REGULAR ☐ SELECTIVE	**Field Change Order** DATE APPROVED	DIV	NO	REV
			PAGE 1 OF		

MFG EFFECTIVITY			REF. ECO	NAME OF PRODUCT AFFECTED	FCO KIT P/N			
PROD. ID NO	ECO S/N	FCO S/N		FIELD EFFECTIVITY (S/N)	ITEMS AFFECTED	YES	NO	
					MANUALS			
					SOFTWARE			ESTIMATED MAN HOURS/UNIT
					SPARE PARTS			

THIS FCO MUST CONTAIN THE FOLLOWING SUBJECTS IN ORDER 1 REASON FOR CHANGE 2 MANUALS AFFECTED 3 DESCRIPTION OF CHANGE 4 LIST REFERENCES 5 CHANGE INSTALLATION PROCEDURE 6 PARTS AND SPECIAL TOOLS REQUIRED 7 REMOVED PARTS DISPOSITION 8 LIST ATTACHED DOCUMENTS 9 SOFTWARE DIAGNOSTICS REQUIRED	1. REASON FOR CHANGE

Originator	Date	Engineering	Date	Mfg	Date	Field Services	Date	CM Mgr	Date

Figure 11.1
Example of a field change order (FCO) form.

2. A part number for the FCO kit such that we can inventory it in our normal parts inventory system.
3. Field effectivity is shown by serial number and again we have a miscellaneous items affected block to ensure that we get the total impact of the field change accomplished.
4. Last but not least, the manufacturing effectivity is shown so that we know what serial numbers are going to be changed in manufacturing for this particular change.

As for the content of the FCO, there are several items which are shown on the left hand side of the form in Figure 11.1. A reason for this field change is no more than a statement of the reason why we need the change. As shown in the list of subjects, which must be shown on the FCO form, we must list the part numbers of all the manuals impacted by the change. Also, a complete description of the change must be done, preferably with marked-up documents, listing any references which must be made for the customer or field service person to use when installing the change, and then, a change installation procedure must be shown so that you can tell the customer or field service person how to install this change. A list of all parts and any special tools required also must be included. One of the more important parts is to list the removed parts and how they are to be disposed of. For example, are they to be returned to the factory or can they be kept by the customer or field service engineer? And what am I to do with my spare parts in the field? Any special documents required for installing this change should be included with the FCO. Last but not least, if this field change involved software then we must give the proper instructions for changing any software. And finally, a final test and check out procedure must be included so that we can test out the system when we have completed the installation of the change. As stated previously, approval of the FCO is the same as for the ECO, which supports the FCO but must also include a field service organization representative.

Behind every FCO there should be at least one ECO document. The FCO form is not used in the same way as an ECO form as it is not the authority in changing engineering documents. It is only used as a directive to install changes in a product which is located at a customer site. Normally, because an FCO can be installed at many customer sites by many different field service engineers, the form must be fully typed. The instructions for installing the FCO must be very explicit and laid out in a very procedural manner. Every FCO not only is approved by the same people who approved the ECO but approval should also include a representative of the field service organization. Once an FCO has been approved, an FCO package or kit is assembled to be sent to the affected customer sites in the field.

11.2 Contents of an FCO Package

The FCO package, or kit, usually includes the following items:

1. A copy of the approved FCO form.
2. An FCO parts kit which includes all of the parts and tools necessary to install the change in the field.
3. A set of updates to the appropriate operating and maintenance manual for the product. These updates can come in any one of the following four ways:

   a) Describe the manual changes in the description area of the FCO documents using a from-to situation.
   b) Include brand new updated pages to be inserted in the manual.
   c) Use marked-up pages from the old manual to show what needs to be changed for this particular change described in the FCO.
   d) Use an addendum to the manual. This is the same as including the new pages but because of the volume of the change or other reasons it is not appropriate to insert new pages into the manual itself and so an addendum would be made to this specific operating or maintenance manual.

If your company should have the appropriate manpower and resources, you may find it necessary to create a new maintenance type database in order to track your maintenance activity in the field in order to know the exact configuration status of a product located at a customer site if you are involved in a formal field change system. This database would track all serial numbers in the field and any maintenance activity which is done against any serial-numbered product in the field so that at any time we would know the exact maintenance status of a given product in the field. This type of information is extremely important if you are to maintain customer satisfaction and to use that information further in any new designs, which might be done so that the same problems do not exist in the new, redesigned product. Though such a database can be very expensive, it is very worthwhile and something like this usually exists in large companies. An example of this kind of activity is what our automobile companies are doing today when they let us know if a particular part of your automobile is in danger of breaking down while it is being driven on the road. We all hear of companies recalling certain automobiles because of certain problems. That's where a maintenance type database becomes extremely important in tracking the maintenance status of our products in the field.

11.3 Summary

A formal field change system is only needed by your company if they find it necessary to make changes in the field. But we have only described an example of a formal field change system. Remember that a field change is only processed if the problem has occurred in a product after the normal manufacturing process, which means usually that the product is located at a customer site. Field changes are processed by the same organization that processes engineering change orders. Approvals of field changes are the same except that the approval of the representative of the field service organization is added. When a cost estimate is generated for the ECO, which is a back up to the FCO, the cost for installing the change in the field should be included in that overall cost estimate. Quality assurance is responsible in the case of a formal field change system for ensuring that all contents of the field change order package are included before shipment to the field. Any further questions on the field system as described in this chapter should be referred directly to the author.

Chapter 12
The Change Organization

12.1 Change Organization Title

The change organization is responsible for handling and controlling all engineering documents in your company as well as for managing the engineering change processing system. The title of this organization will be entirely up to your company, but some of the current titles in industry today are as follows: engineering document control, engineering change administration, configuration management, configuration control, and so on. The author prefers configuration management as it encompasses all of the configuration elements by definition.

12.2 Supervisor Title

The title of the person managing or supervising this organization varies. The author prefers configuration manager or manager of the configuration management department. Other titles may be configuration manager, engineering change administrator, configuration control manager, documentation manager, and so on. Again, manager of the configuration management department is most appropriate. The person managing this department should have the following background:

1. Experience in a technical or manufacturing function.
2. Knowledge of data processing or management information systems.
3. Good communication skills to converse with other company employees and management personnel.
4. Good common sense.

Other people who might be part of the configuration management department include secretary, data entry clerk, drafter, designer, technical writer and anyone required to assist in the overall control and process of engineering documents and changes.

12.3 Change Control Board

If a change control board is used in your company for the processing of engineering changes, that board is usually considered attached to the configuration management organization.

12.4 Configuration Management Staff

The obvious question to ask at this point is where should the configuration management organization be located in your overall company organization? In most companies, you will find that this organization is located within the structure of the engineering organization. Sometimes the manager of the configuration management department reports directly to the executive of engineering, but in other cases and usually in large companies the engineering department has an organization called engineering services. It is within this organization that you will usually find the configuration management organization located. Hence, the manager of the configuration management department would report directly to the manager of engineering services who usually reports to the executive of engineering. A typical company organization chart is shown in Figure 12.1, which depicts the location of the configuration management organization primarily in a large company, usually over 1,000 employees.

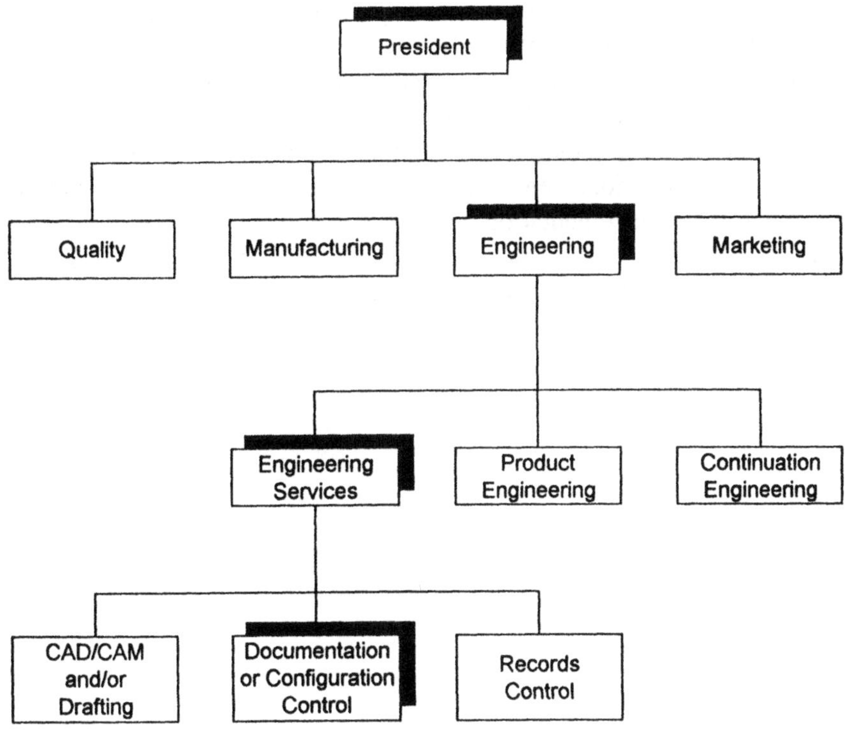

Figure 12.1
Example of a company organization chart.

12.5 Summary

We have discussed the organization responsible for controlling engineering documentation and processing engineering changes. We have discussed the responsibilities of the manager of that department as well as the location of the department in a company organization.

Through my experiences, I know that some companies have this organization reporting to people in departments other than engineering, such as quality, or even directly to the CEO. It is the author's opinion that the interest of engineering and other departments are best served by structuring the configuration management organization within engineering, as they are solely responsible for the product design and that is where all engineering documents are initiated. It is extremely important that people realize, and especially those in engineering, that the documentation being controlled and managed within engineering is documentation that will be serving practically all other functional areas within your company. Engineering must use good common sense when they are providing this effort.

Chapter 13
Configuration Status Accounting

We now will discuss the last major element of a good configuration management or engineering documentation control system, as shown in Figure 3.1, namely, configuration status accounting. This term is defined as follows:

Configuration status accounting is the process of knowing the documentation status of each product by serial number prior to shipment and knowing the product status when it is located in the field.

As stated in the definition, we want to be able to account for the status of the configuration or documentation of our product.

13.1 Basic Information Required

Whatever configuration status accounting system you use in your company, it should provide the following kinds of basic information:

1. Change traceability.

2. Parts usage list (where-used file).
3. Material traceability.
4. As designed/as built verification by part number.
5. As shipped/as current history by serial number.

These activities can be accomplished in many ways and there is no one system that is better than another. The only point is that your configuration status accounting system should include the above information.

13.2 Function Responsibilities

The company functional responsibilities for configuration status accounting are as follows:

1. Engineering is responsible for the design of the product and maintaining up-to-date documentation at all times.
2. Manufacturing is responsible for updating their processes when engineering changes are made and for feeding to the configuration management organization the actual effectivity of each non-interchangeable change which is implemented by date, lot number, or serial number.
3. The field organization is responsible for feeding to the configuration management organization the actual effectivity of each field change by date, lot number, or serial number.
4. The configuration management organization is responsible for publishing and distributing the status of the configuration or documentation of a product by ECO number and product number both internally within the company and possibly externally to the customers as required.

The major status accounting reports required are as follows:

1. Always post the actual effectivity of the ECO and redistribute it once it is changed. This is called "closing the ECO."
2. A bill of material report, by product, which shows changes installed, changes approved by date, lot number, or serial number.

3. A spares part list unless included in the bill of material, with approved changes shown by date, lot number, or serial number.
4. Any other reports as required for your company.

13.3 Main Status Control Document: BOM

The main document which must be kept up-to-date at all times is the bill of material (BOM) for your product. The bill of material will provide the internal product configuration history which is required and includes the part numbers to identify all documentation associated with that product. A bill of material which includes revision letter status of each part number will be extremely hard to keep up-to-date because of the unusual number of interchangeable-type changes which are usually associated with a product. It is almost impossible to keep a bill of material up-to-date which includes revision letters. It is the author's opinion that revision letters do not need to be included in the bill of material if the rules of interchangeability are adhered to. Whether I have revision B of a particular part or revision G of the same part, they both ought to be used interchangeably within my product *if I have adhered to the rules of interchangeability*. Hence, this is an excellent reason why rules of interchangeability are a must in trying to control the configuration or documentation of our products.

13.4 Controlling Product Configuration in the Field

The product identification plate, which usually exists on each product, such as on our home appliances, contains information as to the product name, the product or model number of the appliance and serial number. This information is usually needed in order to obtain parts to repair your particular appliance when it fails. Also, when we ship this device to a customer, we need to know its product or model number and serial number for future reference in case of potential problems.

Some companies use another form, as shown in Figure 13.1, called a product configuration log, if they are using a field change system. The intent of this form is to maintain a history of what was installed in this device when we shipped it from the factory. If there are a number of options which could be a part of this device, we can

PRODUCT CONFIGURATION LOG

BASE PRODUCT IDENTIFICATION NO. ________________ SERIAL NO. ________________ PAGE NO. ____ OF ____

INSTALLED OPTIONS					
PRODUCT ID NO	SERIAL NO.	PRODUCT ID NO.	SERIAL NO.	PRODUCT ID NO.	SERIAL NO.

FIELD HISTORY				
PRODUCT ID NO.	SERIAL NO.	CHANGE NO.	DATE AND WHO INSTALLED	REMARKS

Figure 13.1
Example of a product configuration log.

track the options which were installed in the device at the factory as well as those that may be installed or deleted in the field. Also, on this form we can track any field change activity that we may have against this product. As an example, if we install an FCO package as previously described in Chapter 11, we can record the product number, the serial number and the FCO number on this form. And now, if we should have a potential problem with this device in the field, knowing its product number and serial number from the product identification plate, we can tell what kind of change activity has occurred without doing anything else. Hence, if you have need for a field change system within your company for your products, you may want to consider the use of a product configuration log.

13.5 Summary

Configuration status accounting encompasses maintaining all the records necessary for us to know at any time the current or historical status of the configuration or documentation of a product. Some of the key items which should be included in such a system have been noted in this chapter. Most of this information can be maintained on a database which might be associated with the MRP and bill of material system.

If you have need for further information on configuration status accounting it is suggested that you reference Mil-Std-1521 entitled, "Technical Reviews and Audits for Systems, Equipment and Computer Software." Another reference you might want to look at on this subject is the EIA (Electronic Industry Association) Engineering Bulletin Number CMB6-6, entitled "Configuration Audits."

Chapter 14
Automating a CM or Documentation System

As most of us know, for many years now there has been a growing need for fast and more effective communication within a given company. Early on, back in the 1970's, word processing changed the way that we created and communicated information. Then came the advent of personal computers which changed the way we communicate within an organization or company.

Soon thereafter many industry people started discussing the need for a "paperless" office. The idea was to have information flow electronically from computer to computer and from person to person. The feeling was that information would be more accessible in that printing and paper handling costs would be reduced or even possibly eliminated, and file cabinets would be a thing of the past.

Currently, the paperless office is a dream, not a reality. Although it is possible to share information electronically, the use of online documents did not necessarily improve the quality and the usability of paper documents. The users of this information lost the ability to have text and graphics in a highly-readable form and there was not as much portability of information as was expected. Sharing information among employees was next to impossible because of the use of different brands of personal computers. Since then, many advances in technology have permitted the development and use of several different major software programs which are and can be of great benefit to companies in the process of moving to a paperless system. An example would be the LINKAGE product, which was designed and produced by a company called CIMLINC, located in Illinois. LINKAGE is a product that is so well-conceived that it changes our views and understanding of systems integration. LINKAGE can be thought of as an overlay, which sits on top of existing applications and data sources. LINKAGE contains tools which can be used for a wide variety of forms-based business applications. There are many other types of software packages which can be used for the same purposes as LINKAGE and some of those packages are described in Appendix A5.

14.1 ISO 9000: The Computer Integrated Manufacturing (CIM) Connection

Chances are that your company has sought, is seeking, or has received ISO 9000 certification. For those not familiar with ISO 9000, it is a set of international standards for quality management. These standards and their associated certification process state:

That your company should be governed by a comprehensive, well-documented, and up-to-date set of procedures;
That these procedures should be followed consistently throughout your entire company;
That corrective actions should be taken whenever quality problems or procedural violations do occur;
That procedures and practices should be updated as conditions occur or need to be changed.

The complete set of what we have referred to collectively as the ISO 9000 standards are composed of five different standards. The ISO 9000 standard is a guide to the selection and use of the four documents listed below:

ISO 9001 is the most comprehensive standard and covers the entire process of product design through field service.

ISO 9002 describes the production and installation procedures for a product.

ISO 9003 applies only to final inspection and testing of a product.

ISO 9004 contains general guidelines for the development of a company's quality management system. These guidelines are quite generic and say very little about the control of specific processes, levels of quality, or defect rates. There primary focus is on documentation and adherence to controlled procedures. Specifically, ISO 9004-7 is now the hardware configuration management document.

The ISO standards, though very comprehensive, appear to be weak in their attention to information systems and technology and in the management of computer-based documentation. The current version of ISO 9000 standards is under review and it is expected that future additions will be stronger and more direct in their treatment of computer-based systems.

The author of this book was deeply involved in trying to determine if a certain multi-division, multi-national, multi-billion dollar company was adhering to the ISO 9000 series of standards of quality management, and if in fact the procedures did cover all the elements of the ISO 9000 standards. By comparing the company's existing procedures to those required by the ISO 9000 standards, we were able to determine their adherence to having a good quality system based on the ISO 9000 standards.

If you are considering organizing or revising your current or new engineering documentation control system, you are urged to review all of your procedures against the ISO 9000 series of standards before you start so that you will fully understand whether you are in compliance with these international standards on quality management.

14.2 Use of Online Documentation vs. Paper

In the past, the use of paper has had some distinct advantages, such as:

Compound documents. Paper can offer an excellent mix of text and graphic information to improve readability and increase understanding. Currently, there are online document systems which do provide the benefits of text and graphic information that users have come to expect from paper.

Readability. Given the choice of identical paper and online documents, most readers will choose paper because of its readability. But the readability advantage of paper over electronic documents is shrinking. Computer hardware is being offered which improves the screen resolution and sophisticated automatic-type scaling technology. Document authors are learning how to improve online readability with relatively simple formatting changes. Greater use of white space and larger type sizes can help close the readability gap between paper and online documents.

Portability. As most of us know, paper offers almost an unmatched portability as it can be distributed and used virtually anywhere; newspapers and paperback books are excellent examples of this. But new compact portable computers and view-only display systems are moving in the right direction in overcoming this problem with electronic documents.

14.3 Cost Savings

The savings associated with reducing printing and paper cost have often been cited as a primary reason for a paperless office. But that does not necessarily include everything because some of the hidden costs of paper are usually quite staggering, such as the cost for mailing, storage and filing. This would easily outweigh the original cost of printing the materials.

For documents which require periodic updates, online documentation is very appropriate. Online distribution of this information will have its cost significantly reduced because of this effort. Use of a compact disk (CD), magnetic tape, a floppy disk, or other disk-based products can significantly reduce your cost in comparison to the use of paper.

14.4 Evaluating Online Systems

Evaluating online systems can be a very complex and confusing task. There are many different products, architectures, and product features available and sorting out their capabilities and marketing promises makes the task even more difficult. Here are a few ideas to help you in this area.

14.4.1 Evaluating Your Application

Your move to having online documentation should not be driven simply by a desire to use new technology but rather by quantifiable good business reasons, such as will it save time, will it save money, and will it improve the usefulness of the information.

14.4.2 Understand the Sources of Your Online Documents

Consider how your information will be prepared for online distributions. Will existing documents be put online or will we have to create documents? Is there a large base of existing documentation that needs to be put online? If documents are already available electronically can they be converted into the format required for the proposed online system?

A new strategy that can be used for converting to online documentation is to begin by putting only newly-created documents online. At some point, most current documentation will be available for online viewing but it is critical that all documents eventually be put online if that is the direction we are going. There is a method for converting these documents: Service bureaus can usually scan the documents via OCR (optical character recognition) to transfer them into electronic form. Do your documents need to be edited, updated or reused? Very little information is created once and never edited, updated or reused. Will new documents need to be added to the online set of documents? There are many situations where several people will want to add or update online information. Hence, we should look for systems that make it easy to add new documents to our collection without having to do any rework.

14.4.3 Evaluate How Online Documents Will Be Used

You should try to understand how and where all of our online documents will be used and which are necessary to be included in our overall documentation system.

14.5 Online Document Review/Approval

The process flow of online documentation is really not much different than what the old paper flow was, or is; but review of the documents is much faster as well as much more accurate. One of the major processes that should be in place in your company is concurrent engineering; where engineering or design allows the other functions, such as manufacturing, to be involved in the original design of the product. Concurrent engineering, as stated earlier, is the same as the product design team. The main goal of concurrent engineering is to get the best features included in the design the first time so that the product is not only manufacturable but is reproducible at a reasonable cost to the customer. With the use of concurrent engineering, if it is working as it should, the review/approval cycle of documents should be much quicker than if engineering and manufacturing groups were working separately.

Related to concurrent engineering is having the supplier of any of your parts be involved as part of the design or approval cycle. This would be ideal. His or her input would be from the reproducibility viewpoint as he or she is probably the expert. As far as the design review of a product is concerned, it should be done in parallel between engineering and manufacturing such that all parties present, including engineering, manufacturing, and the supplier, can discuss any revisions or changes to the design.

The next level of review would be to insure that each individual part does indeed fit with its mating part within the total product design. This activity is called design verification and is a very valuable part of the entire review and approval process. It verifies that the parts fit together and are aligned correctly on the CAD screen through simulation and analysis before your supplier actually goes out and builds the real part as directed by the associated documentation. With the introduction of CAD systems, parts can be designed and verified immediately in the 3-D display on the CAD screen to insure proper fit and alignment. In the past we have had to wait for the prototype to be built to see design flaws. Now the CAD database can do it much better than the hard prototype because the CAD system is working with exact dimensions using simulation and analysis.

14.6 Need for CAD Standards

A set of organized standards is needed for controlling the documentation generated by your CAD system. Some of the standards that should be required are as follows:

CAD and 3-D methodology . A standard on this subject will define requirements for using a CAD system to design and document hardware electronically. This standard would insure the commonality of product documentation generated on a CAD system within your company.

CAD system access. This standard should describe the access and security requirements for a CAD/CAM computer system as security is an extremely important requirement within the product design documentation.

2-D and 3-D drawings. This standard would define the practices for generating 2-D and 3-D drawings using a given CAD control system. This standard would cover all of the requirements for the various types of drawings that you would be generating in your company.

Leveling and line weights conventions. This standard would define which levels of a CAD-generated drawing shall be used for storing information on specific aspects of items in assembly drawings and the suggested line weights to be used for plotting those drawings with a CAD system. The same database is used or should be used to create all types of drawings, hence, a level standard is required to provide a standardized method to facilitate interdepartmental usage of drawings within your company. A level is defined as a preassigned layer in the CAD system for specific categorized information.

Attributes. In order for information, such as part rotation, shape type, etc., to be made available in your CAD applications, attributes and semi-attributes must be assigned to define certain entities. For definitions of attributes and entities, see Appendix A3.

Checking. This organization or function would define the requirements for what must be checked on design documentation which has been generated electronically. Since the CAD-generated design documentation is stored in a database, it is the documentation master and must be verified so that the database can be used and revised accurately and efficiently.

CAD drawing format. This standard would establish the minimum requirements for only a CAD-generated drawing format. Some of the items which may have to be delineated in this standard are the title block format, the revision record block, and any restrictive legend that is needed on the document because of your company's business. It is suggested that existing industry (ANSI) and government (Mil-Stds.) standards be used to assist in this activity.

Glossary. This would only be a standard that contains a list of the terms used throughout the company's CAD standards.

14.7 Summary

We have covered a great deal of information on the automation of engineering documents without getting involved in a lot of detail. The reason for this presentation is because there are so many different automated document systems available today that it is difficult to specify one system over another. My reference to the use of the LINKAGE program available through CIMLINC is because it is an excellent database which can assist you in the automation of your documentation. The kind of CAD workstation you choose will certainly depend on the kind of documentation you have and how to best control it. I have only attempted in this chapter to provide you with a framework for establishing a system for automating your documentation.

Remember, the design group should develop all parts, bought or made, within your CAD system and should perform a design verification process such that manufacturing and the supplier can work within the database for the purposes of design, manufacturing, releasing the parts, and placing the parts on order. The CAD system must have a database manager smart enough to notify the affected people when these items are ready for design review or need updating per your particular engineering change system. They will then have their review decision available to all concerned. The system has to be smart enough to ensure proper adherence to document control and then it forces discipline for the users to follow a set process.

Remember that we have discussed many details of automating documentation throughout this book. Below is a list of some of the key items discussed with the proper section number indicated:

Title	Section No.
Paperless Systems	1.5
Needs for Using Automation	1.6
Use of Computer Simulation and Modeling	2.5, 5.1.3
Use of CAD and MRP Systems	2.5
Use of Significant vs. Non-Significant Part Numbers as Related to Automation	4.4.1
Parts List on CAD-Generated Drawings	4.6.1
Controlling CAD Drawings and their Revisions	4.9
Use Change Control Philosophies When Switching from Manual to Automated System of Documentation	6.5
Automated Change Control Forms	7.2
Automated Change Processing System	10.0
Configuration Status Accounting Database	13.1

With the above table and the other information in this chapter and Chapters 15, 16, and 17, you should be able to follow the points which should be covered in automating your documentation for the future.

Acknowledgement

Information contained in this chapter is the combined result of material obtained from: G.L. Bartuli, Consultant, St. Paul, MN; CIMLINC, Inc., Itasca, IL; Frame Technology, San Jose, CA; and C.W. Lieske, Ceridian Corp., Minneapolis, MN.

Chapter 15
Guidelines for the Transition to a Configuration Management or Documentation Database

In today's world, many companies have a configuration management database, or are in the process of developing such a database, or possibly the company is working strictly with a manual configuration management system. Many databases are available in the private sector offered by companies who have developed the necessary software for such a database, as shown in Appendix A5. What we will discuss in this chapter are some of the key items that you should look for when considering and developing a configuration management database for your company.

15.1 Benefits of Automation

Most of us in industry today seem to understand what kind of benefits we will achieve by automating a certain aspect of our business. Generally, these benefits consist of the following items:

A. Data for the system is captured at the source of the information.
B. Information that is produced by the system is much more accurate than what we have been able to achieve in a manual system.
C. There is improved information quality because of the consistency of the information within the database.
D. Use of a database will insure that procedures are consistently executed.
E. Information stored in the database can be shared by many individuals, departments and functions within a company and everybody receives the same information output.
F. The use of a database will provide management with an information base that can be relied on within the company in order to graph trends in documentation and to make any future decisions on changing your processes.

These benefits are generally true for any database conceived and developed for a particular functional area within your company. Hence, by having a transition from a manual system to an automated system, we should be able to achieve all of the benefits as described above.

15.2 Establish the Scope and Requirements for the Transition from a Manual to Automated Documentation System

We must emphasize here that we "must plan ahead" such that we do not have a system that becomes obsolete the first time it is fired up.

The scope of the system we want to develop should incorporate the goals which we want to achieve for our company. Some of these goals should be as follows:

The system should have file control within it such that we have a CAD library.

It contains or interfaces with the bill of material system within our company and provides the design function with an online parts list generation capability.

The system contains all of the controls needed for a strong configuration management system, including the three major items of identification, control, and status accounting, and should be the prime source for providing electronic engineering change generation.

Our configuration management system must be able to integrate with any existing MRP system.

It must be able to satisfy the needs for a group technology environment within manufacturing in order to enable design location of parts and assemblies.

The configuration management system should be integrated with the financial system within the company in order to provide a closed loop of design information to support the customer and manufacturing (to keep invoices and product serial numbers synchronized).

Software engineering development should be a part of this configuration management database system so that computer-aided software engineering can generate their needed documentation.

The project management and status must be controlled within the system and integrated with the company's project review system.

In trying to define the data and system requirements for this database, it is imperative that we :

1. Include subject matter experts, that is, people who are experts (actual "doers") on specific areas involved in the database.
2. Ensure that we consider what the users want from this database.
3. Ensure that we comply with any industry or contract requirements for our kind of business.
4. Develop and utilize prototypes for the various major sections of the database in order to provide a fantastic opportunity for subject matter experts to experience computer technology and to make any recommendations.

15.3 An Architectural Sketch

Figure 15.1 is an example of the general architecture which should be included in a configuration management database. As you can see, we have all the elements for the as-designed condition of a product being fit into a master CM database. The as-planned condition, as noted here, also is included in the central database and from this information we are able to produce the as-built condition and henceforth the as-shipped condition of the product.

15.4 Existing Data Management Systems

Appendix A5 contains a listing of a number of configuration management software programs and packages. This is a very dynamic list. For more up-to-date information, contact the author.

15.5 The Selection Process: Yours or Mine?

In trying to establish our database requirements, we must *understand* the various needed hardware and software items and their limitations in order to determine what hardware or software is required for your particular configuration management database. Then, we must determine what our various goals are as far as the operation of the system is concerned. In the hardware area, we may want to consider the use of personal computers, workstations, a departmental computer system with server and last but not least, a large central server to operate the entire system. In the area of software, we have to understand relational database management systems and the pros and cons of an object-oriented database management system. Are we going to use third or fourth generation language in our system? Can we make use of artificial intelligence and expert systems within our database system? What are the advantages of a UNIX operating system for our workstation? What kind of editors and word processors are we going to have? Many questions will occur during this selection process. One of the main questions is, Are we going to purchase the hardware or software or are we going to try to design and build it ourselves?

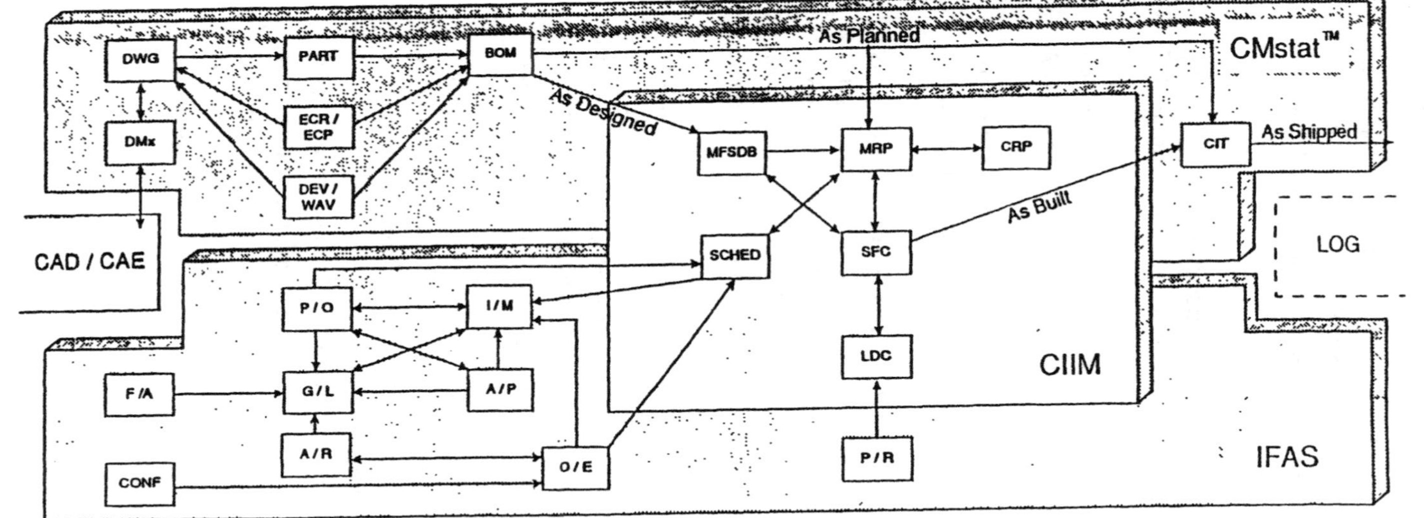

Figure 15.1

Architectural sketch for a CM database. Key: BOM: bill of materials, CIT: configuration item tracking, DEV/WAV: deviation/waiver, DM_x: file control, DWG: documentation management, ECR/ECP: change management, PART: part/BOM/routing information, CONF: configuration, CRP: capacity requirements planning, LDC: labor data collection/bar code, MFSDB: manufacturing standards/database, MRP: materials requirements planning, SCHED: master scheduling, SFC: shop floor control, A/P: accounts payable, A/R: accounts receivable, F/A: fixed assets, G/L: general ledger, I/M: inventory management, O/E: order entry, P/O: purchase order, P/R: payroll/personnel management, CM stat™: engineering information management system, CIIM: computer interactive integrated manufacturing, IFAS: interactive financial accounting system. Source: E.I.M.S., Inc. and Digital Information Systems.

Hence, we create what we can call the vendor system vs. the home grown system. Some of the considerations given to whether we purchase or build our own configuration management database system are shown in the list below.

1. Is writing software one of the primary businesses in your company or can a vendor provide the software for us more economically? (What are your priorities?)
2. If there are few hardware or software products available which meet our requirements, can we get a better database by developing our own? (Probably not!)
3. Will we benefit from existing vendor user groups from our vendors when we ourselves do not have similar experiences? (Probably yes!)
4. Cost is a primary factor in any new database development. It is not unusual for a home grown system to cost between 1 and 5 million dollars, and they don't always work effectively. Whereas a vendor can possibly provide software which would range in cost anywhere from $1,000 to $350,000. (Depends on how much uniqueness and a perfect fit are worth!)
5. Even if a company perceives that their processes are unique, it is possible by reviewing the existing vendor systems they will find an existing turn-key configuration management system. (Is uniqueness always better? Attendees in the author's seminars generally agree that they are not really unique they are just different terminology!)
6. A home grown system will have maintenance which is a large factor over a system's life time. Whereas with the use of a vendor system, they can amortize the cost over a large base of users. (It is usually difficult to get a support budget!)
7. The technology included in the database in a home grown system will be directly proportional to the resources available within your company. Whereas vendors will have access to experts in the state-of-the-art technology. (Software suppliers will provide advance code to application vendors to use on their next update because of their large installed base!)

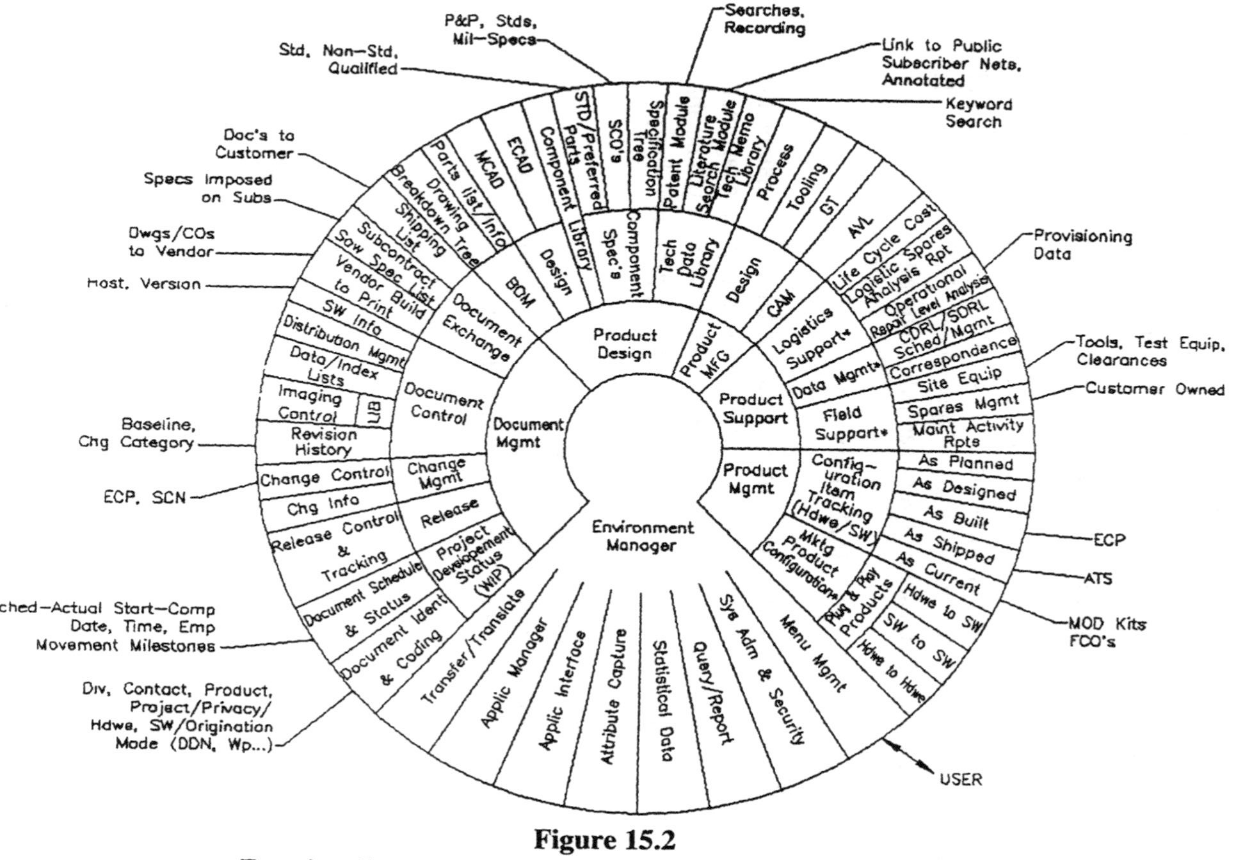

Figure 15.2
Functionality template: The whole thing. [CM wheel, Source: Grayme Bartuli (1991).]

8. With the proper configuration management database system we will be able to identify what we a) are suppose to build!, b) are building!, c) have built!, hence we can support it.

15.6 Functionality Template: The Whole Thing

Figure 15.2 shows a very comprehensive diagram of a good configuration management database. It is safe to say that you could write a book about the components of this wheel. But, I believe, it is self-explanatory if you study it very carefully yourself.

15.7 Summary

We have tried to provide some general guidelines for making the transition from a manual to an automated configuration management system. If you follow the elements of this chapter, you should be able to evolve a configuration management system appropriate for your company.

Acknowledgement

Some of the information contained in the chapter is from material obtained from G.L. Bartuli, Consultant, St. Paul, MN.

Chapter 16
How and Where to Start Creating or Revitalizing a Company's Engineering Documentation Control System

This chapter will deal with the necessary ingredients for creating or revitalizing a company's engineering documentation control system, or possibly you call it your configuration management system. In either case, the ingredients are the same. In the case of revitalizing an existing documentation system, you may find that some of the ingredients shown here already exist within your particular system. If so, you do not have to make any changes to your system as you will be matching your existing documentation system with the ingredients discussed in this chapter. In order to make it easier for the reader, we have chosen to break up the ingredients for creating or revitalizing a company's engineering documentation control system into five major phases. They are as follows:

Organization and planning.
Items required for a configuration identification system.
Items required for a configuration and change control system.

Miscellaneous items required for a company's engineering documentation control system.
Overcoming the management barrier to CM.

We will now discuss the content of each item but first we have to know certain things about our company before we start. Some of these items are as follows:

1. What is our kind of business?
2. What are our company's goals and strategy?
3. What are the engineering and manufacturing plans for producing a product in our company?
4. Are we going to have a product design team in our company as discussed in Section 2.3?
5. Do we know our competition and customers? Are we regulated by any industry or government agencies, e.g., FDA, FCC, DOD, or NASA?
6. What kind of engineering documentation do our customers expect and how does our competition control their engineering documentation?
7. Has the marketing group in our company determined the product numbering system that we are to have and what is our plan for maintaining our product in the field?
8. Has the manufacturing group determined whether they are going to use assembly processes or drawings or both in assembling the product in their area?
9. Have we identified a person to lead the engineering documentation control effort in our company? Using this checklist, the lead documentation person should propose a structured framework for our new or revised engineering documentation control system based on the content of this book. Their recommendations should include the various documentation types required in our company for our kind of business; should we be considering a manual, semi-automated or a fully automated documentation control system? What kind of forms should we be using in our system? What formats should be required for our system, such as schematics, drawing and specification format?

We will now review in detail each of the five phases described above.

16.1 Organization and Planning

Two things should happen before we get into the detail of creating or revitalizing an engineering documentation control system. First, an experienced person should have been selected to lead the engineering documentation program. Their initial job would be to propose a structured framework for an engineering documentation control system with the various recommendations required to have a good system. Secondly, form a committee composed of members from the major functional areas of your company, including any key people that should be a part of the process of creating or revitalizing your company's engineering documentation control system. Typically, the functional areas represented on this committee would come from the following groups, marketing, engineering, software, manufacturing, manufacturing engineering, production control, quality, purchasing, engineering records control, and any other group which should be represented in this effort. Do not forget user representation. This committee should be essentially the same as the product design team, already discussed in this book. The chairperson should be designated by the engineering executive of your company and should have been trained in how to run a committee. Meetings of this committee should be held as often as required by the work involved -- every day, once or twice a week, or once a month -- depending on the level of activity. The full effort of the committee should be to lay out a plan of action which is documented and defines the purpose, goals, objectives, and action items required for each phase of developing our company's engineering documentation control system. Each action item should include a date of approximate completion. The action items should relate to the various items required under each phase of activity.

16.2 Items Required for a Configuration Identification System

The following items are the minimum required for a good configuration identification system and hence, consideration should be given to projecting out the needed ingredients for 25-50 years:

1. Define the types, formats, forms and the required approvals which will be required for the engineering documentation control program. Some of these types, formats, forms, and approvals will be for drawings, specifications, engineering changes, and so on, whatever your requirements might be for your company's business. Remember, we should use a minimum number of approvals, keep things simple and not complex, and have a user friendly system.
2. Define the various numbering systems required to control the various kinds of documentation needed in your system. As stated previously, you should have a minimum number of identifiers, such as a part number, a serial number, a change number, and as previously stated, a product numbering system as described and controlled by your marketing group.
3. Start assembling the terms required for a glossary for your documentation system.
4. Company procedures or standards, including any flowcharts to define needed procedures, may be required to document the necessary rules for generating a drawing or specification either manually or on a CAD or word processing system. Most rules for generating a drawing can be found in a set of drafting standards from the American National Standards Institute (ANSI), 1430 Broadway, New York, NY 10017. If these drafting standards are used, there would be no need to have your own company standard on this subject. Standards or procedures may also be necessary to describe your release system, part number system and your serial number system. Examples of the formats which may be required for your system can also be found in ANSI standards or in some of the illustrations shown in this book.
5. Develop a simple documentation release system with easy recognition and usability.
6. Evolve a company policy on your documentation or configuration management system for enforcement purposes.

16.3 Items Required for a Configuration and Change Control System

As we are now discussing configuration and change control, the following items should be included in the development of this phase:

1. Use of a set of interchangeability rules.
2. What is the philosophy of change for our company?
3. What change forms are we going to use? How many approvals?
4. Will we use an engineering change request form in our system or will the ECO form act as an ECR also?
5. Do we have a need for an engineering change cost estimate form in our company?
6. Do we have need for a deviation and/or waiver process in our company?
7. How will we process or analyze a change?
8. Are we going to make the bill of material the prime configuration control document? If so, how?
9. How are we going to control the maintenance of our product in the field? Do we need a field change system?
10. Can we use 3-D simulation to reveal potential change problems before the fact?
11. Is our system going to insure that we get the *total* impact of a change?
12. Evolve a company policy on the total dollar value for approving a change.
13. How are we going to control and determine spares in our company?

16.4 Miscellaneous Items Required for a Company's Engineering Documentation Control System

Some of the items which should be considered for an overall engineering documentation control system are listed below:

1. Obtain a cage code, if required, for our kind of business. This code is obtained by contacting the Defense Logistic Center in Battle Creek, Michigan.

2. Evolve a CM ingredients chart similar to the one depicted in this book.
3. Evolve a classification and coding system, if needed, to work with your newly developed part number system if it is non-significant and for a group technology effort in manufacturing.
4. Evolve a engineering documentation control training program.
5. Evolve task groups of specialists under the documentation committee which was formed to evolve a new or revised engineering documentation control system. Each task group will have a specific task to complete, such as determining what kind of part number system will be used in your company.
6. Consider the need for having a CM plan in your system.

16.5 Overcoming the Management Barrier to CM

For years, engineering documentation people have struggled with getting management support for their major problems, such as:

> Change control board meetings which are not accomplishing anything because of personnel not performing their responsibilities.
>
> Everyone only considering documentation as a necessary evil and hence, very little attention is given to it by management and others.
>
> Documentation people are not given the proper respect and support they rightly deserve.
>
> Documentation rules and procedures are not adhered to by company personnel.

These are only a few of the problems confronting documentation people. The list could go on and on.

So what can we do to overcome these and other problems which are a detriment to a good engineering documentation control or configuration management system?

It is the author's strong belief that company management must get more deeply involved in the engineering documentation control process. As stated in Chapter 18, I believe that having a brief company policy on the major elements of your company's engineering documentation control or configuration management system is one of the most important steps in order to make company and management personnel accountable for their actions in the documentation control process. A company policy also should be required for delineating the control of costs of engineering changes, such as what level of management can approve what total cost figure for the total impact of an engineering change.

In addition, copies of all procedures and flowcharts which describe the documentation control process should be readily available for *all* involved management and company personnel. The procedures and flowcharts, of course, can also be accessible in an online documentation control system. Paper copies are *not* a necessary evil.

Last but not least, most companies have been exposed to some type of total quality management system as defined by the ISO 9000 series of standards as well as the 14 quality management principles defined by Dr. W. Edwards Deming, a well-known quality expert in this country.

Again, it is the author's belief that to have a good company quality management system, it must contain the necessary procedures, etc., for a good engineering documentation control system based on the principles in this book. And, per Dr. Deming's 14 management principles, the Engineering Documentation Control System must be adhered to by all company personnel as defined by company policies and procedures.

So what are Dr. Deming's 14 management principles? They are as follows:

1. Create constancy of purpose toward improvement of product and service with the aim of becoming competitive, staying in business, and providing jobs.
2. Adopt the new philosophy.
3. Cease dependence on inspection to achieve quality.
4. Cease doing business on the basis of price tag alone. Instead, minimize total cost.
5. Improve constantly and forever the system of production and service.
6. Institute training on the job.

7. Institute leadership. The aim of leadership should be to help people do a better job. Leadership of management is in need of overhaul as well as leadership of production workers.
8. Drive out fear so that everyone may work effectively.
9. Break down barriers between departments.
10. Eliminate slogans, exhortations and targets.
11. a) Eliminate numerical quotas on the factory floor, substitute leadership.
    b) Eliminate management by objective, eliminate management by numbers, substitute leadership.
12. a) Remove barriers that rob the hourly worker of his right to pride of workmanship.
    b) Remove barriers that rob people in management of their right to pride of workmanship.
13. Institute a program of education and self-improvement.
14. Put everybody in the company to work to accomplish the transformation.

As a friend of mine once said, "Companies are in business to make a profit -- not to lose money." We must have *all* company personnel working together in order to have a successful and cost-effective engineering documentation control or configuration management system. Barriers created by management and others will certainly hinder this process.

In summary, if we all follow the golden rule of "do unto others as you would have them do unto you," we will all be happy and successful in our documentation activities. There is no magic formula to a successful documentation control program except for trying to "Act smarter rather than harder" and invoking the old KISS principle (Keep it simple, stupid). Hopefully, the principles described in this book will make things a little easier for each of you.

16.6 Summary

We have tried to delineate an outline from which a company can develop or revise their company engineering documentation control system. We start out by organizing and planning for this activity by determining our purpose, goal, objectives and activities required.

Management barriers to a good engineering documentation control system are also discussed. Most companies should be able to formulate a good company engineering documentation control system by using the outline contained in this chapter as well as all the information previously contained in this book.

Chapter 17
The Future of Configuration Management in the Private Sector

As we look into the future of engineering documentation within a configuration management system, we will try to forecast what direction we in private industry will be going as far as controlling our engineering documentation. With the new computer technology, it is readily apparent that we will be handling our documentation differently in the future.

In the future, computers and computer software, using advanced database transfer techniques, will play a major role in every aspect of controlling our engineering documentation. Skilled professional documentation personnel will be dealing with versatile, sophisticated products whose characteristics are a blend of hardware and software technology. With most design and modifications being performed on interactive computer terminals, it will be very difficult to tell where software ends and hardware begins. Like it or not, configuration managers will require software management skills -- even to manage the hardware.

Unfortunately, we won't be able to throw away all our pencils and paper, but advanced computer technology will absorb many present day mechanical repetitive functions. Most of our configuration-related data will be software-generated by-products. As configuration managers are freed from most of the clerical aspects of documentation control and with complete and precise information available to them, the configuration manager will play a very important role in program management decision-making.

The challenge to today's configuration management practitioners is to look beyond limits of today's techniques or risk becoming anachronistic, as the fabled bookkeeper with the green eye shade and quill pen.

Our world is changing dramatically. What I think we can see, if we really look hard enough, is that some of the things on which we are focusing a great deal of attention on today are becoming obsolete before our very eyes. While we concern ourselves with, among other things, engineering reports, specifications, types and forms of drawings, deferred data procurement, and even engineering change documents, there is a distinct possibility that for a large segment of private industry, these and other present forms of data may not exist in the future.

The accelerating advance of technology devices produces today things unheard of just a few years ago. Today, I carry in my shirt pocket, or even hold on the tip of my finger, the computing capacity that only yesterday filled an entire building. Design and manufacturing methods are evolving with the use and assist of the computer. Computer-aided design and manufacturing is here today, it is maturing, and is rapidly becoming the way of life in many of our industries. Throughout all of this evolution there is the ever present, increasingly powerful, and always changing technology of the computer. Computing hardware is getting smaller and cheaper, and each year we are doing more with the software than we ever thought possible.

17.1 Functional Models Supplementing Specifications

Now let us gaze into the proverbial crystal ball to examine the potential technological evolution in configuration management likely to occur in the future as a result of, and in conjunction with, the expanded capabilities expected in computers and software.

I will attempt now to describe how automated techniques may replace much of today's product documentation; how automation will assist the change decision process; and how status accounting in real time will act as a downward driver of our costs.

17.1.1 Automated Product Documentation

As we look into our crystal ball, we see functional computer program models supplementing and, in some cases, completely supplanting paper versions of well known configuration identification documentation, such as system product and material process specifications; test plans and procedures; and engineering drawings. These mathematical models, representing the real time configuration of the product to be produced, are developed initially for conceptual phase studies and are continuously refined throughout subsequent phases of the product development. They are created using a general purpose modeling system which can be used to refine and incrementally expand the mathematical model.

An interrelated hierarchy of these models is created and refined as the requirements for a system evolve and are allocated to lower levels. Product definition and description suitable for analysis, production, maintenance, modification, and reprocurement are created by computer-aided design data processing techniques. Most parts are produced using computer-aided manufacturing techniques. At any instant in the process the deliverable hardware and software of any given configuration are directly traceable to software stored in computer memory. All associated documentation is in the form of stored intelligence, which is accessible in readable form from computer terminals, CRTs, and occasionally, in paper or microfilm form from printer or plotters.

The models are used to develop test procedures and to evaluate changes to the design through simulation. They provide interaction through computer-to-computer cross talk between and among the associate contractors, sub-contractors, equipment suppliers, and other facilities.

The deliverable data at the end of the process may completely definitize the model containing the total product description in the form of a portable data file rather than in hard copy. It represents all the data needed for maintenance overhaul, logistics, and reprocurement and yes, it is small enough to fit in my shirt pocket.

Documentation as we know it today will be minimal in the future. Some percentage of the equipment of the '80's and '90's is still a part of the functional inventory and has to be supported with techniques that are a mixture of what we do today and partially implemented 1990 methods. For new products, methods of design and management are consistently evolving, paced only by the state-of-the-art in processors and peripheral devices.

17.1.2 Using an Automated Change Control Board (CCB)

As we continue looking into our crystal ball, we are now picking up what seems to be a change control board meeting. At the head of the table is a projection screen which is linked to a computer terminal. When a question is asked on a change problem, a technician keys it into the terminal and the answer is immediately projected on to the screen. Automation has now entered the change control arena. This is not a frill; functional design changes can be so rapidly effected that, by necessity, change decisions have to made very quickly. Otherwise control of the product configuration is lost, along with consequent cost, inventory control and profit. Present day paper work cycles will be intolerable at this time.

Automation in the change control and change board decision-making process involves the exercising of mathematical models by computer programs which perform change analysis in evaluation. It also includes interfaces of databases containing inventory information status and work process records. Each change is simulated by playing it through the model prior to its implementation to determine, almost instantaneously, its total impact, including the effect on performance, compatibility, interfaces, logistics, reliability, maintainability, and true life cycle costs.

Failures or deficiencies in production, field failures, and customer-directed actions serve as preliminary initiators for impending changes. Field failure, reliability, and maintainability information is computer-collected, analyzed and provided in summarized form for change action when predetermined thresholds are approached.

Design activities respond by exercising mathematical models to determine potential fixes. Engineers entered the detail design into an uncontrolled computer field in which trial and error changes will not permanently affect stored designed data. Once the preliminary engineering solutions are reached, configuration management convenes the change board for the decision process, and the computer links and audio-visual aids allow real time evaluation of the change.

Computer-stored designed data, including preliminary design change considerations will be simultaneously in view for all CCB members, including remotely-located members who are tied into the process, possibly by telephone/video/computer consoles.

Many alternate design solutions, including the choice of *not* making the change, are analyzed to determine their impact. Each change is subjected to a cost benefit trade off analysis. Often the change that is on the surface does not appear to be the best solution and it ends up being the choice because simulation shows it to have the best cost performance ranking over the entire life cycle.

17.1.3 Computer-Aided Change Implementation

After change approval and authorization, the computer again will play a part in the implementation of the change. As we just described, preliminary changes made for CCB evaluation were accomplished by entering needed data into an engineering control portion of the computers "revision segment." The initial design remains undisturbed until change approval is received, at which time revision segment data will be brought in to modify the base file.

Performance and requirement parameter changes will be permanently inserted into the identification model or product specification file. As an example, output voltage to the specification will automatically update the corresponding voltage in lower tier requirements.

Instructions for direct operation of machines, rework or replacement orders, and updates to manufacturing plans, work schedules, and "cruel" loading charts will be initiated.

Parts changes will cause a computer search for the corresponding parts listed in drawings, operation, maintenance, and parts break down handbooks or manuals, provisioning lists, and test procedures, all resident and related databases. Parts changes in the computer will also trigger procurement, stocking, and eventual issuance of changed items.

17.2 Status Accounting Comes of Age

With only a slight adjustment to the focus control of our crystal ball console, we see that configuration status accounting has finally come of age. It brings the "real world" of product operation, maintenance, and logistic support in close accord with configuration identification and change control. Once again computer technology makes a major contribution. A solid database of configuration data delineates the real time status of all items in the inventory. Through computer interrogation, it provides the change control board, as well as the logistic and maintenance planners, with the facts and analytical tools for accurate evaluations, cost productions, and decisions.

As a result of advances in state-of-the-art database cross talk, and other remarkable methods of data communication, a closer vendor-customer-using activity team relationship has developed. Gone are the misunderstandings caused by the lack of information and interpretive forecasting tools. The conflicts among program participants are now minimized because the improved data banks accommodate dynamic inquiries and provide accurate and incisive responses that are timely and complete. Early problem identification and enhanced visibility of current status force relationships to become more honest and open. As a by-product, they tend to highlight any area of data overlap and redundancy that can subsequently be eliminated by mutual consent.

The 1990 status accounting systems are now designed to capture and transmit reliable information by remote means while still providing security control for sensitive data.

Input techniques are not rigidly standardized. They allow individual methodology and procedures for loading the database, procedures which are extremely flexible and adaptable to a wide variety of contracting and product circumstances. What is standardized, to a large degree, are the common data elements being entered into the automated data banks and extracted by users at either local or remote locations. The new technology has fostered data element standardization at the same time as format flexibility.

The key change status data and as-designed configuration data are created as by-products of the automated design and change control processes and are integrated with other management procedures. Status data, schedules and lead times are automatically calculated at every stage of the change process. From request for change through CCB schedules, decisions, customer approval if required; and then release of the updated design from uncontrolled to active status in the machine memory. Very little paper is being generated except for action-oriented reports. Key milestones are automatically interjected into program management's schedules and work authorizations. Decisions and implementing orders are conveyed to those in need of the information and, where required, automatic stop effort notices are produced for areas whose in process work is impacted by pending changes.

Status accounting as-built input methods cover the spectrum, from handwritten punch cards to techniques such as optical scanners that monitor hardware configuration and provide online updates. Some of the newer software oriented equipment contains within its memory a status accounting record of its own specific configuration. This configuration record is remotely accessed and modified as changes are implemented or as parts are replaced by maintenance actions. And these later circumstances, total traceability of configuration from the shop floor through the using activities operational environment, can be accomplished by integrating the appropriate database. The keys to this traceability are machine assigned unique serial number identifications or signatures of all the parts and assemblies down to the replaceable part level. Specific signatures for software versions are also used to control the configuration of highly modularized software products.

17.3 Exercising Control

These new computer techniques enable hardware and software to be controlled throughout the development, production, operational, and maintenance phases in a continuous coordinated fashion. Change activity triggers a closed loop system which follows a change until physically incorporated in the product, whether in production or by retrofit by the user of the product.

An automated search of current as-built data records provides an immediate and completely accurate assessment of the configuration of delivered products or systems. The optimum serial number for both production and retrofit of a change can thus be determined at the change control board meeting, as well as a quantitative analysis of the number, type, and installation point of any modification kits required.

Field modification data and kits parts list are by-products of revision of the basic design data and will be generated by the computer models. Dependent upon the capability and equipment at this scheduled installation site, the instructions may be transmitted by hard copy or directly to remote computer terminals.

As modification kits are assembled and then issued, as-built data is appropriately annotated to provide an automated suspense record. Modification kits include computer coded forms, cards, chips, or marked sensitive devices used to feed back incorporation data. The field installer checks the completion block and returns the data to update the file either by prepaid mailer or by accessing a remote reader tied to the computer network. Since the incorporation action is scheduled via an interfacing automated system, a "tickler" is produced if the required response is not received in time, and "past due" notices and delinquency reports are sent to the installing activity and appropriate command stations after a suitable grace period. This system functions, when used, to control periodic maintenance and other time compliance requirements.

The volume of information that is compiled in this database network is immense. Yet each user will be able to extract the information important to him/her in real time and in the format most usable to that person. The new flexible reporting techniques allow us to ask the computer for any explicit set of related elements, and a logical analysis of that data. We can, for example, ask for trend charts, statistical studies, financial summaries and the like, and we can get our answers almost instantly. The enormous memory capacity in ultra-high speed processing of future computing hardware makes these and other computers extremely inexpensive. The commitment to online systems in visual displays makes "zero based" paper flow a practical reality.

17.4 How Do We Get from Here to There?

Now that we have described what some may call a configuration manager's utopia, let us come back to the present and look at some of the practical realities. Is this future really possible?

Well, in some large, advanced private organizations there already exists systems which have begun to approximate components of the general system that we have described. We have already spoken about the current advances of computer-aided design and manufacturing. Some automation of the change control process has also begun at several companies, and several have developed significantly automated status accounting and verification systems. The growth of these concepts and their acceptance by corporate management over the next few years is, of course, greatly dependent on the rate of advancement and the subsequent application of new computer technology and resources.

Economics, as always, will be the primary driving force behind a gradual evolution effected by industry as a whole. As present day engineering and management personnel are gradually replaced by college graduates whose whole educational process has been influenced by the computer, the acceptance of widespread computer use in all areas of business will be tremendously accelerated. The new breed of manager will have been "weaned" on the computer and will not have the mistrust that seemingly still prevails among many in top level management.

Will the new technology leave the little guy out in the rain while the giant corporations play there computer games? I don't believe so. Current cost trends in processors, peripheral equipment, and leased line microwave communication networks strongly imply the accessibility of the new techniques to businesses of all sizes.

As far as configuration management managers go, it does not take a genius to realize that we had better get software-oriented and learn how to manage software -- but fast! While the new "support " software systems are being created for automated design and manufacturing of hardware, and "software factory" techniques are applied for the development of software, we must take an active part. Unless our systems of identification and control of configuration are built into the support software, we will not be able to keep pace.

If we continue doing the job as we are today despite technological advances, we will force excessive and unnecessary costs and excessive work to be layered on top of the basic in machine design production efforts. The new computer-minded managers will be quick to see the inefficiency of the "paper pushers."

17.5 Summary

We see a new frontier in configuration and data management. Our challenge is to define it and apply our innovation and imagination to the solution of its problems. We must evolve with the times by using the technology of tomorrow to manage the products of tomorrow. If we don't, we are in danger of becoming obsolete.

The message is clear. We must promote today the discipline, tools, and caliber of personnel required to manage tomorrow's sophisticated world of configuration and data management.

Acknowledgement

Information contained in this chapter is the combined result of material obtained from G.L. Bartuli, Consultant, St. Paul, MN and the Electronic Industries Association G-33 Committee on Configuration Management.

Chapter 18
Overall Summary

We have now covered the four major parts of a good configuration management or engineering documentation control system, as shown in Figure 18.1. They are configuration planning, configuration identification, configuration control, and configuration status accounting. Each one of these elements is a very important part of your overall documentation system. We must be able to plan our documentation in advance of starting our product design, we must be able to identify all the various elements of our documentation including the parts, the assemblies, the specifications and whatever other documentation is needed for your kind of business. We must be able to control the documentation against all the future changes which might be made against that documentation. And last but not least, we must be able to account for the status of the configuration and documentation for that product. As stated in the opening section of this book, it is imperative that we include all of these four major elements in any good engineering documentation control system. Hence, we must be certain that our system is well-documented in the form of either engineering practices or procedures using flowcharts, and we must ensure adherence to them by all employees.

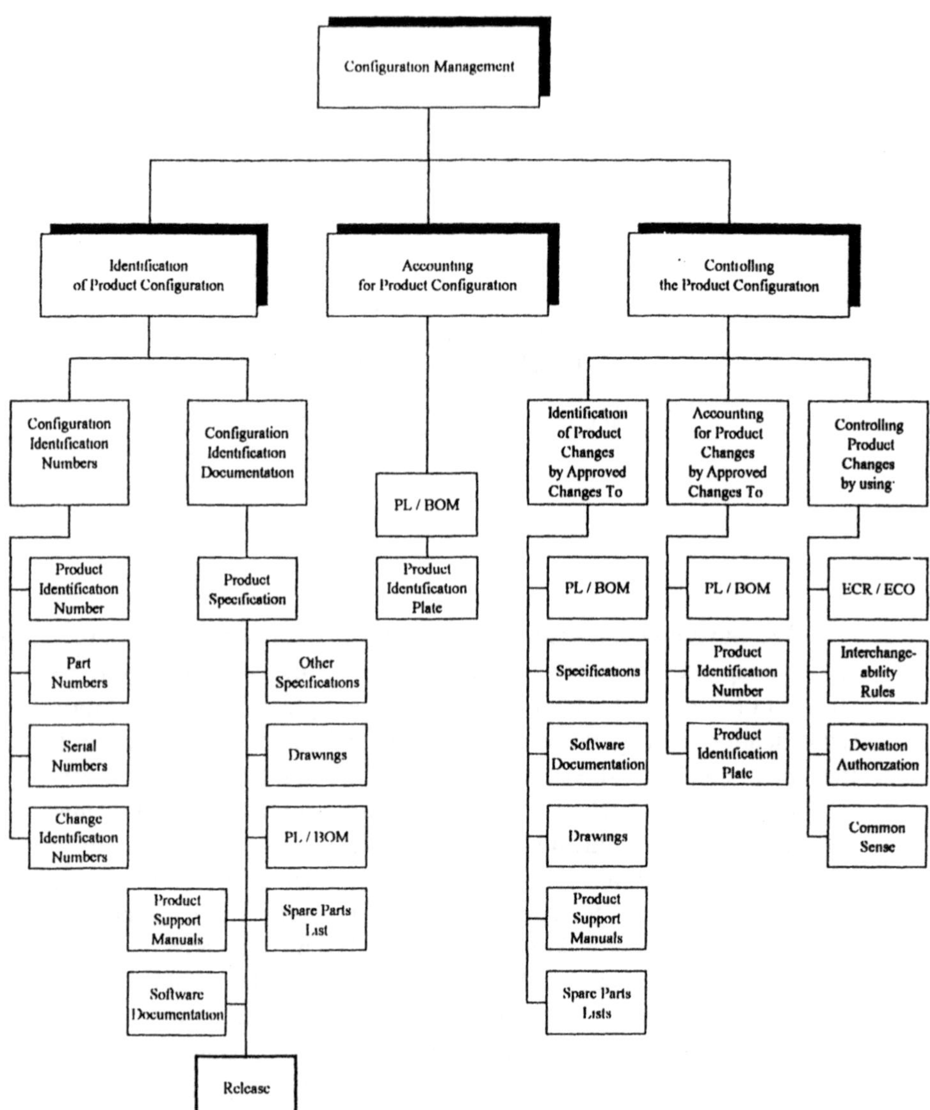

Figure 18.1
Ingredients of configuration management.

In Figure 18.2, a typical company policy is shown which can be used to enforce the use of the ingredients of a good configuration management system. Note that it will be company policy to have an organized and consistent management process for controlling the configuration or documentation of our hardware and software products and that process shall be used by all development and manufacturing operations throughout a product's total life cycle. Because people working in the documentation area have a lot of problems in keeping the discipline within a good documentation system, management support is vitally needed. Usually a company policy on this subject is one way of obtaining that management support and getting the discipline necessary to ensure having a good engineering documentation control system.

As we conclude this book, you should have found information that will be of help to you either in developing or revitalizing your current documentation system. I cannot impress upon you enough the necessity for including a representative of the users of your system in any kind of committee that must suggest possible changes to your documentation system. The user is extremely important and is usually forgotten when we start forming committees to review our documentation systems. It is important that a configuration management system is implemented early on, in the conceptual phase of the development of a product. The use of a product design team hopefully will insure that happening.

A lot of companies in today's business world are hoping some day to achieve some kind of a paperless system in their facilities, but it has been proven in some research studies that a full paperless system will be nigh to impossible in the oncoming years. A typical paperless system might be as shown in Figure 18.3. Though we have tremendous capabilities available to us today in data processing systems being offered by industry, it is quite often found that it is almost impossible to get along without paper. The most difficult part usually is in the approval process of a drawing generated at a CAD workstation or on an approval of an engineering change. Also, there are a number of other problems associated with trying to establish a totally paperless system. But whatever we can do in this area certainly will be a benefit and will assist in expediting the generation of new drawings or specifications and even future changes to these documents.

PURPOSE

This policy defines a set of management disciplines for utilization in the planning, identification, control and accounting of a product's configuration from its inception till its obsolescence. By using these disciplines, the company will be driving for the lowest total life cycle cost, insuring the stated product performance, providing adequate logistic support and insuring the compatibility of hardware and software products.

POLICY

It is the policy of the company to have an organized and consistent management process for controlling the configuration of hardware and software products. This process shall be used by all development/manufacturing operations throughout a product's total life cycle.

This management process is known as Configuration Management and consists of Configuration Planning, Configuration Identification, Configuration Control and Configuration Status Accounting. The Configuration Management process for hardware and software products is described in other company procedures.

GLOSSARY

Configuration management is the systematic approach to planning, identifying, controlling and accounting for the status of a product/system configuration from the time it is conceived throughout its whole intended life. Thus it provides information on what we have built, are building, are marketing, are changing and are supporting.

Configuration planning is the process of determining how a product shall be configured (documented).

Configuration identification is the technical identification and documentation used to define the approved configuration of a product.

Configuration control is the : 1) process for controlling changes to a configured item during its life cycle; 2) procedures for documenting and reporting change processing and verification; and 3) external relationships required to maintain system compatibility.

Configuration status accounting is the process of tracking the configuration of a product from the time it is shipped and throughout its entire supported life.

RESPONSIBILITY

The senior executive of the company is responsible for the implementation of this policy.

Figure 18.2

Typical company policy on configuration management.

So how can we best control the configuration and documentation of our products? I would suggest that you follow the following eight most important items:

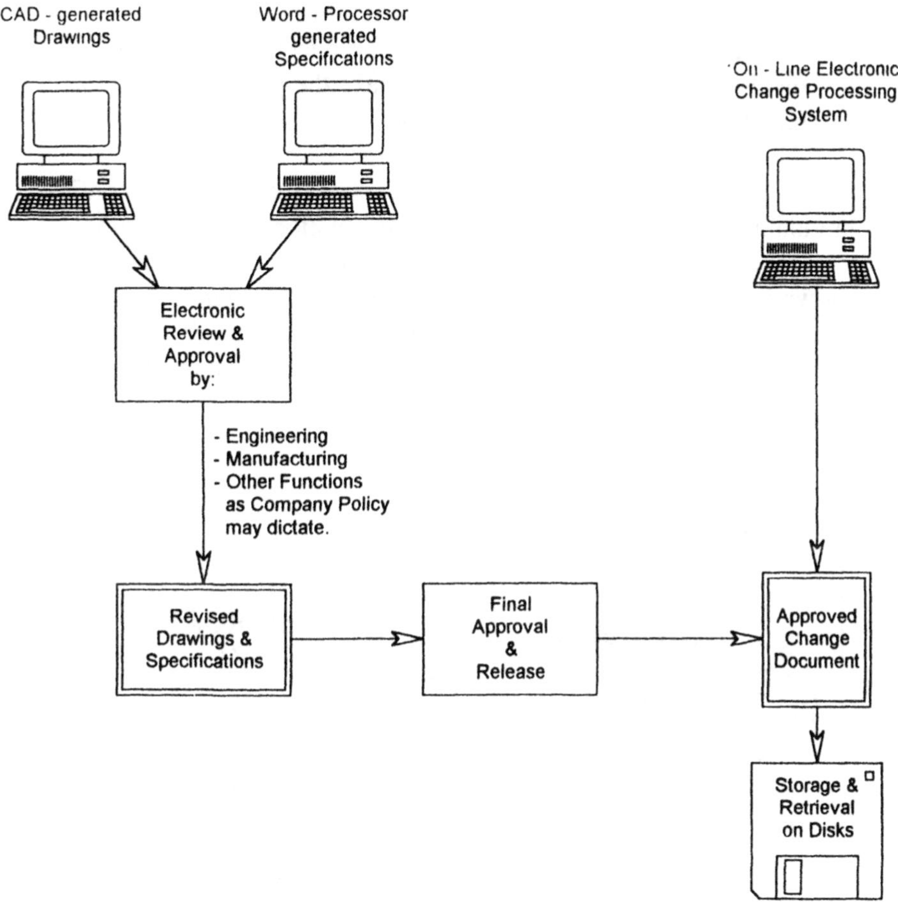

Figure 18.3
Example of a totally paperless system.

1. Use your company procedures and follow the rules contained therein without trying to go around "left end."
2. Manufacture products which will be shipped to a customer only by fully approved and released documentation.
3. Document *all* changes required no matter how small.
4. Use the interchangeability rules properly.
5. Use serial number or lot effectivity for all Class I changes.
6. Keep your bill of material updated at all times.
7. Once a serial number is assigned to a product never change it.
8. Field changes are only required to make the product perform and function per its product specification.

These are the items which must be adhered to for a good configuration management or engineering documentation control system.

I hope you have found this book interesting and that maybe, as shown in Figure 18.4, you will be able to clean up your act, clean up your engineering documentation control system and get on the road to a very successful program of engineering documentation control.

CLEANED UP
THE ACT

Figure 18.4
Cleaning up your act.

Appendices A1 - A5

Appendix A1: Interchangeability Rules

Glossary

1. Interchangeability, interchangeable, interchangeable item - When two or more parts possess such functional and physical characteristics as to be equivalent in performance and durability, capable of being exchanged without alteration of the parts themselves or any adjoining parts except for normal adjustment, and do not require selection for fit or performance, the parts are interchangeable.
2. Non-interchangeable. When a part is to be changed and the changed part does not physically fit or function as a replacement for the unchanged part in all applications of its previous revisions, and all unchanged parts will not physically fit or function as a replacement for the changed part, the parts are considered non-interchangeable.
3. Service level. Service level is that point of assembly within a product which has been established as the level at which problem diagnosis or repair or both will take place in the field. Below this level, assembly components are neither repaired or spared unless customer requirements specify otherwise.

Requirements

1. General

1.1 Each item (i.e., product, sub-assembly, component, and detail part thereof) shall be identified by a part number which shall be, or shall include, the drawing number. The revision letter may be included in the identification, but does not serve as an identifier for non-interchangeability.

1.2 Drawing and part numbers are assigned in accordance with company policy.

1.3 Rules.

1.3.1 When an item is changed, and the change is of such a nature that direct and complete interchangeability is maintained, the item part number need not be changed, but its drawing revision letter is advanced such as, A to B, B to C, etc.

1.3.2 When an item has been changed in such a manner that superseded and superseding items are not directly and completely interchangeable, the item part number shall be changed.

1.3.3 The part number shall not be changed when a new usage is found for an existing part.

1.3.4 Exception. In the event that the changed item will physically fit and function in place of the unchanged item, but the unchanged item cannot be used in place of the changed item, the change is considered non-interchangeable and will require an assignment of a new part number to the changed item. However, if there are no products in the field and if all of the unchanged items are either scrapped or reworked to the new configuration, the change may be made by advancing the revision letter.

1.3.5 While normally an interchangeable change would result in only a revision letter advancement, the serviceability requirements may require that a new part number be issued for the changed item. For example, a new part number must be assigned if:

1. Items in the field must be replaced because of reliability or safety reasons.
2. Items in spares inventory must be purged to assure change on failure.

1.3.6 Reworked items. When an existing item is reworked to be directly and completely interchangeable with a new item, it shall be identified by the part number of the new item.

1.3.7 Changes to items which require new part numbers must be reflected in the next assembly by either a revision letter change to that assembly or new part number for that assembly. The part numbers of all progressively higher assemblies need be changed only up to the assembly where interchangeability is preserved.

1.3.8 Changes to Printed Wiring Board (PWB) Assemblies. If the PWB assembly is reworked for a non-interchangeable changed, the part number must change of the reworked PWB assembly. The part number of the PWB itself will not change. Subsequently, if the reworked PWB assembly becomes a new

PWB assembly via an update of its documentation by an ECO, the part number of the PWB itself must change but only the revision letter need be updated of the part number of the reworked PWB assembly assuming no additional change to the PWB assembly.

2. Service Level and Serviceability

2.1 The primary factor of consideration in establishing service level shall be the serviceability at that level. Once service level is established, a minimum requirement to acceptable serviceability shall be that the product is in fact serviceable and maintainable at and above the service level. Furthermore, any change order written against that product shall apply the rules of interchangeability at the established service level.

2.2 For a given product, service level may vary from the major assembly level to the discrete component level.

Example No. 1. Service level of the power supply may be determined to be at the power supply assembly level, dictating that once the problem is diagnosed to be in the supply, then the entire power supply subassembly is replaced.

Example No. 2. Service level on a terminal board assembly may be determined to be at the discrete component level. In this instance, the problem must be diagnosed to be a defective component and that component is replaced.

Appendix A2: Acronyms

ANSI	American National Standards Institute
ASSY	Assembly
BOM	Bill of Material
BSI	British Standards Institute
CAD	Computer-Aided Design
CAGE	Commercial and Government Entity
CAM	Computer-Aided Manufacturing
CB	Competitive Benchmarking
CCB	Configuration/Change Control Board
CI	Configuration Item
CIM	Computer Integrated Manufacturing
CM	Configuration Management
CSA	Canadian Standards Association
DWG	Drawing
ENGRG	Engineering
ECO	Engineering Change Order
ECP	Engineering Change Proposal
ECR	Engineering Change Request
FCO	Field Change Order
GT	Group Technology
HW	Hardware
IC	Interchangeable Change
JIT	Just-In-Time
MFG	Manufacturing
Mil	Military
MOD	Model or Modification
MRP	Material Resource or Requirements Planning
MTBF	Mean Time Between Failure
NIC	Non-Interchangeable Change
PL or P/L	Parts List
PO	Purchase Order
PWA	Printed Wiring Assembly
PWB	Printed Wiring Board
QA	Quality Assurance
SCD	Specification Control Drawing
SCO	Software Change Order

SCP	Software Change Proposal
SN or S/N	Serial Number
SPEC	Specification
SQC	Statistical Quality Control
STD	Standard
SW	Software
TLA	Top Level Assembly
UL	Underwriter's Laboratories
WCM	World Class Manufacturing

Appendix A3: Glossary

Approved Engineering Document	A controlled engineering document which has been approved by engineering, at a minimum, and by other departmental functions as your company policy may dictate.
Assembly	A number of parts or sub-assemblies or any combination thereof joined together to perform a specific function and capable of disassembly.
Baseline	An approved reference point, at a specific time, for control of a product's design, development, manufacture, and maintenance.
Bill of Material	A complete listing of all piece parts, assemblies and any other items which are required to build and ship a product to a customer. A bill of material encompasses all parts lists required in a product plus any other required items for shipping the product such as product support manuals.
Breadboard Model	The assembly of hardware representative of a product or a portion thereof used for experimental and testing purposes as well as proving out a new design by engineering.

Cage Code	Commercial and government entity Code (formerly FSCM).
Change	A modification to a product which is necessary to meet the requirements of the product functional specification, or to meet safety standards, or to reduce manufacturing or maintenance costs.
Change Number	The number that appears on the ECO form which is used to track changes to a product.
Class I Change	See Non-interchangeability.
Class II Change	See Interchangeability.
Component	A part, assembly or product which is purchased from an outside vendor. Examples of components are fasteners, resistors, integrated circuits, clamps, power supplies, etc.
Concurrent Engineering	See Product Design Team.
Configuration	The technical description and arrangement of parts or assemblies or any combination of these which are capable of fulfilling the fit, form or functional requirements defined by the applicable product specification and drawings.

Configuration Control	A systematic or organized method for identifying, controlling, and accounting for the status of changes which affect the configuration or documentation of a product.
Configuration Identification	The technical documentation which is properly identified and defines the approved configuration of products under design, development, and test, in production or in the field.
Configuration Management	A discipline for providing a systematic or organized approach to planning, identifying, controlling, and accounting for the status of a product's configuration from its inception throughout its intended life (that is, from birth to death).
Configuration Management Plan	The document which describes the detailed requirements for identification, control, status accounting and audits necessary to manage the configuration of a given product.
Configuration Planning	The process of determining how a product shall be configured, documented, and supported.
Configuration Status Accounting	The process of knowing the documentation status of each product by serial number prior to shipment to a customer and

knowing what the product status is when it is located in the field.

Controlled Engineering Document	One that is under document identification control which consists in the assignment of an initial unique document identifier (part number), imprinted on the document, and revised in accordance with the rules of interchangeability.
Deviation	A short term or temporary departure from an approved engineering standard, specification, drawing or other engineering document of a company designed and manufactured or purchased product effective on one or more of its serial numbers.
Documentation	That paper work needed to technically define and support products and systems designed and manufactured by your company.
Drawing	An engineering document that discloses by means of pictorial or textual presentations, or combination of both, the physical and functional requirements of an item (DOD-Std.-100).
Effectivity	Determining exactly when a change will be or was implemented in a product on the manufacturing assembly line.

Engineering Change Cost Estimate	A form used to accumulate the total cost of a Class I (non-interchangeable) change.
Engineering Change Order	A form which describes an approved engineering change to a product and is the *authority* or *directive* to implement the change into the product during the manufacturing process.
Engineering Change Package	All of the ingredients required to process and implement an approved ECO including the ECO form, all affected marked-up documents, and any other document as may be required by your company.
Engineering Change Request	A form which is available to any employee in your company to use in trying to describe a proposed change or problem which may exist on a given product.
Engineering Document	A drawing, specification, process, artwork, parts list, bill of material, or other type of document which originates in the engineering department and relates to the design, procurement, manufacture, test or inspection of items oı services as specified or directed therein.
Engineering Release	The use of a controlled engineering document which is

needed for limited procurement of parts and assemblies on breadboard or prototype models.

Field Change Order (FCO)	The directive to install changes in a product after the normal manufacturing process in order that the product will perform to its written or implied specification, to meet safety standards, to improve reliability, or to reduce maintenance costs.
FCO Package	All of the ingredients required to process and implement an approved FCO in the field, including the FCO form, parts, tools (if any) to install the change, manual updates, and any other documents as may be required by your company.
Full Release	The use of fully-approved engineering documents in the procurement and manufacture of production models.
Inactive Part	A part or assembly not used in a new product design or new production but active in spares inventory for maintenance of existing products in the field. It does not infer any obsolescence of spares or prohibit procurement for spares inventory.
Interchangeability	Occurs when two or more parts possess such functional and

physical characteristics as to be equivalent in performance, durability and is capable of being exchanged without alteration of the parts themselves or of adjoining parts except for adjustment and without selection for fit or performance.

Maintainability	See Serviceability.
Modification	A Class I change to a product after its first shipment to a customer.
Non-interchangeability	Occurs when a part is to be changed and the changed part will not physically fit or function as a replacement for the unchanged part in *all* applications of its previous revisions, and all unchanged parts will not physically fit or function as a replacement for the changed part.
Obsolete Part	A part or assembly no longer stocked, used in design or field maintenance, or used anywhere else in your company.
Option	A marketing term which is a modification or add-on to a product.
Part	One or two or more pieces joined together which are not normally subject to disassembly without destruction of designed

	used. (Examples: resistor, screw, gear, capacitor.)
Part Number	An identifier which uniquely identifies a specific item and is used to control assembly and replacement of assembly parts on the basis of interchangeability.
Parts List	A listing of only those items which are needed to build an assembly in the family tree or bill of material of a product.
Pre-production Model	The complete assembly of a product by manufacturing and not engineering which is built from at least pre-released engineering documentation during an initial production or pre-production phase to check out production techniques. During this phase, production tooling is fabricated, processes are developed and checked, acceptance tests are run and final product refinements are made in the design.
Pre-release	The use of a controlled engineering document which authorizes manufacturing to proceed with production of a specified number of units, to be called pre-production models, provided these engineering documents have also been approved at least

	by the engineering and manufacturing functions.
Printed Wiring Assembly (PWA)	The completely assembled functional unit including the printed wiring board (PWB), components, connectors, test points, etc.
Printed Wiring Board (PWB)	An insulating panel in which completely processed electrical conductor paths are deposited, etched, or printed in various patterns to provide or complete a circuit between two or more points. It may provide contacts for plug-in mounting facilities.
Product	Any item which is produced and offered for sale or lease by your company.
Product Configuration Log	A form associated with the product that some companies use to maintain a history record of what options were installed in a product at the factory prior to shipment and which options were installed or deleted in the field. Also, it contains a listing of FCO's installed in the product.
Product Design Team	A group of people assembled to increase the interaction and communication between the various departmental functions of a company in the design, de-

velopment, and manufacture of a new product.

Product Life Cycle — The time and activity encompassed from birth to death of a product, this is, from the time it is conceived throughout its intended life.

Product Specification — A document which describes what a product is and what it does by defining its physical and functional characteristics.

Production Model — A product which is built on the production line from fully approved and released documentation.

Prototype Model — The complete assembly of a product used for design testing in determining the adequacy of the total design. This product is built under the direction and control of engineering and not manufacturing.

Provisioning — The process of determining the types, quantity and stockpoint locations of parts to be spared (i.e., spare parts, special tools and test equipment) and required to support and maintain a product when in service or use.

Purchased Item	An item which is procured outside the company from an independent procurement source.
Record Change Only	A change made to correct non-configuration affecting errors on the document or to add a tab to an engineering document. It must not affect fit, form or function.
Regular Field Change	A change in which all serially numbered units of the same product configuration in the field are impacted.
Release	The act of supplying valid and fully approved documented technical information on engineering designed products to both procurement and manufacturing.
Reliability Non-interchangeability	Occurs when a part or assembly is changed to improve the reliability to the level necessary to meet the intent of the product specification.
Repair Part	Any individual component part or assembly required for the repair of a repairable part below the service level. (Also known as repair component parts.)
Repairable Part	A part which can be returned to usable operating condition by

appropriate repair action when that action is cost justifiable.

Selective Field Change — An FCO addressed to only units of a specific product configuration which may be installed only when required for a given customer application.

Serial Number — A number used with a product identification number or part number to denote a specific end item.

Serviceability — The degree to which a product can be maintained in the field using the product support manuals for supporting field repair activity at or above the service level.

Service Level — That level on the branch of the family tree or bill of material of a product where it is most economical or practical to spare or service a part or assembly.

Spare Part — An individual item (such as piece part or assembly) which is normally stocked as a replacement for a failed or worn out item of a delivered product to a customer.

Specification — A document used for procurement, design, or manufacturing which clearly and accurately describes the essential requirements for items, materials, or services including the minimum procedures necessary to determine compliance with the stated requirements.

Specification Control Drawing	A document which specifies the needed requirements of a component for your company such that it can be purchased from an outside vendor.
Standard	A document which contains a statement of detailed requirements established by general consent as the most practical and appropriate current solution to a recurring or new problem.
Supply Item	Material required by a customer to allow system operation. Typical supply items are reels of magnetic tape, printer paper, disks, paper tape, etc.
Support Item	Any unique special tools or test equipment required for maintenance of the product being provisioned.
Tabulated Document	A document such as an engineering specification or drawing which encompasses more than one part number. Tabulated documents provide an economical means for describing similar items in a single document in-

stead of preparing separate documents for each item.

Top Level Assembly Drawing	The highest engineering document describing the product purchased by a customer as shown in Figure 4.4.
Waiver	A long-term or permanent exemption from the requirements of a product design standard, procedural or process standard, specification, drawing or any other engineering document used in a company designed, manufactured, or purchased product.

Appendix A4: Engineering Documentation Bibliography

Bills of Material: Structured for Excellence

By: Dave Garwood

An excellent and complete guide and reference manual on bills of material, including simple examples showing how to build them.

Available from: Dogwood Publishing
P.O. Box 28755
Atlanta, GA 30358-0755

Principles of Configuration Management

By: M.A. Daniels

A comprehensive overview of principles with examples, illustration, and bibliography. Government contractor perspective.

Available from: AACI
M.A. Daniels
VSE Corporation
2760 Eisenhower Avenue
Alexandria, VA 22314-4587
(703) 329-1633

Configuration Management Deskbook

By: W.V. Eggerman

All government and military compliance requirements are covered.

Available from: AACI
Tab Books
Blue Ridge Rd.
Summit, PA 17294-0840
1/800/822-8158

Graphical Displays for Engineering Documentation

By: Daniel L. Ryan

Presents a highly organized and simplified approach to solving several complex problems facing engineers, designers, and draftspersons.

Available from: Marcel Dekker, Inc.
270 Madison Ave.
New York, NY 10016
(212) 696-9000

Software Configuration Management
By: Ron Berlack
Hard cover, 392 pages, Published 9/91
Available from: John Wiley & Sons
1 Wiley Dr.
Somerset, NJ 08875
(800) 225-5945

Military Standards

Mil-Std.-12 Abbreviations for Use on Drawings, Specifications, Standards, and Technical Documents

Mil-Std.-973 Configuration Management

Department of Defense (DOD) Standards

DOD-Std.-100 Engineering Drawing Practices

Military Specifications

Mil-Std.-1000 Drawings, Engineering and Associated Lists

Department of Defense Directives

DOD 5010.19 Configuration Management

International Standards

9000 Quality Management and Quality Assurance Standards -- Guidelines for Selection and Use

9001 Quality Systems -- Model for Quality Assurance in Design/Development, Production, Installation, and Servicing

9002 Quality Systems -- Model for Quality Assurance in Production and Installation

9003 Quality Systems -- Model for Quality Assurance in Final Inspection and Test

9004 Quality Management and Quality System Elements -- Guidelines

The United States has adopted these standards and they are available from ANSI as ANSI/ASQC Q90 Series of Standards.

ANSI Standards

ANSI/IEEE Std 828-Software Configuration Management Plans

For Copies of ISO and ANSI Standards, contact:

American National Standards Institute (ANSI)
1430 Broadway
New York, NY 10017

EIA Standards

EIA CMB4-1A Configuration Management Definitions for Digital Computer Programs

EIA-CMB4-4 Configuration Change Control for Digital Computer Programs

Note: The above Mil-Stds. and DOD documents are available through the U.S. Government Printing Office in Washington, D.C., or by contacting one of these groups:

Global Engineering	1/800/854-7179
National Standards Association	1/800/638-8094
Information Handling Services	1/800/241-7824
Electronic Industries Association	1/202/457-4900
Document Center	1/415/591-7600

Appendix A5: CIM Applications

Compiled by Grayme Bartuli, 12685 Dodd Ct., Rosemount, MN 55068, (612) 423-4486. This information is subject to update and does not constitute an endorsement of the products listed. Mr. Bartuli is a CM Data Base Consultant.

Configuration Management Systems

CM System

Bloodhound **Bloodhound Configuration Manager**

Hdwe is:	PC	Woolsey ,Brian ,
DBMS is:		Mgr, Logistics Marketing
OPSys is:	MS/DOS	Thomson-CSF Sys Canada In
$ Range:	$5K Canadian	350 Sparks Street
Partner:		Ottawa Ontar K1R7S-8
Mfr:	Thomps-CSF	Mfg Phone:(
		Distributor:(613)594-8822x

Has Ident, Control, Status Actg, Reviews & Audits. Meets DoD 480, 482, 483 & 2167

Does not have a BOM

C-GATE **CACI's Gateway**

Hdwe is:		Kaplan ,Lisa
DBMS is:		Product Manager
OPSys is:	UnixDOS	CACI, Inc.*
$ Range:		1100 No. Glebe Road
Partner:		Arlington VA 22201-
Mfr:	CACI	Mfg Phone:(
	1-(800)84C-GATE	Distributor:(703)841-4102x

A 3 Phase process to integrate modules & legacy databases into a (custom?) solution.

Exhibitor at CM Conference in Orlando

CADI **Computer Aided Design (CAD) Integration Version 2**

Hdwe is:	MainF	
DBMS is:		
OPSys is:		IBM Corp.
$ Range:		
Partner:		
Mfr:	IBM	Mfg Phone:(

Generate, assemble & extract BOMs & related data from CATIA & CADAM models.

If used w/other IBM CIM products helps manage product/process info from Eng inception to mfg rel.

CCS	**Change Control System & Change Order Processing**	
Hdwe is:		Natzic ,Walter ,
DBMS is:		President & CEO
OPSys is:		Arron-Ross Corp.
$ Range:		1132 Indian Springs Dr.
Partner:	Tandem	Glendora CA 91740-
Mfr:	ARRON/ROSS	Mfg Phone:(818)963-4119
		Distributor:(818)963-4119x

Includes On-Line ECO's, Dev/Wav, Real-Time ECR, ECO status, history & traceability

I haven't seen this module.

CLIP	**Config & Logistics Info Program**	
Hdwe is:		Kaplan ,Lisa
DBMS is:		Product Manager
OPSys is:		CACI, Inc.*
$ Range:		1100 No. Glebe Road
Partner:		Arlington VA 22201-
Mfr:	CACI	Mfg Phone:(
		Distributor:(703)841-4102x

A CALS compliant front-end to CM and Logistics data.

CM-Plus	**Configuration and Data Management**	
Hdwe is:		Natzic ,Walter ,
DBMS is:		President & CEO
OPSys is:		Arron-Ross Corp.
$ Range:		1132 Indian Springs Dr.
Partner:	Tandem	Glendora CA 91740-
Mfr:	ARRON/ROSS	Mfg Phone:(818)963-4119
		Distributor:(818)963-4119x

CMS **Configuration Management System**

Hdwe is:		Natzic ,Walter ,
DBMS is:		President & CEO
OPSys is:		Arron-Ross Corp.
$ Range:		1132 Indian Springs Dr.
Partner:	Tandem	Glendora CA 91740-
Mfr:	ARRON/ROSS	Mfg Phone:(818)963-4119
		Distributor:(818)963-4119x

Includes WBS, Doc/Image Mgmt, BOS, BOM, Effectivities, Rev Status & Vendor Hist

CMsoft

Hdwe is:		
DBMS is:		
OPSys is:		Telavic Israel
$ Range:		
Partner:		
Mfr:		Mfg Phone:(

See TCC under CM Systems

CMstat **CMstat**

Hdwe is:	Multi	Gain ,Paul ,R
DBMS is:	Ingres	Pres, CEO & Founder
OPSys is:	Multi	CMstat Corp. *
$ Range:	$4 - 200K	5755 Oberlin Drive
Partner:		San Diego CA 92121-
Mfr:	CMstat	Mfg Phone:(619)552-6660
	1-(800)927-7828	Distributor:(619)552-6663x

Complete CM system with File Control (DMx), BOM & Parts mgmt features.

CMstat runs on DOS, UNIX, DEC, IBM operating systems

DataCac **DataCache**

Hdwe is:	Multi	Natzic ,Walter ,
DBMS is:		President & CEO
OPSys is:	Unix	Arron-Ross Corp.
$ Range:	$30K +	1132 Indian Springs Dr.
Partner:		Glendora CA 91740-
Mfr:	ARRON/ROSS	Mfg Phone:(818)963-4119
		Distributor:(818)963-4119x

DataCache is the suite of integrated applications for cooperative

processing & data transfer.
Runs on over 500 platforms

DCS **Document Control System**

Hdwe is: PC
DBMS is: Propriet
OPSys is: Personal Computing Tools
$ Range: $349 17419 Farley Road
Partner: Los Gatos CA 95030-
Mfr: PCT Mfg Phone:(
1-(800)767-6728 Distributor:(800)767-6228x

Dwg Rel, Current Dwg Status, Dwg Info, ECN Status, ECN Info, Affected Dwg Info.
Not sure if the "company" is a PC software clearing house or the author of the SW.

ECMS **Engineering Configuration Management System**

Hdwe is: MainF Norton ,Gloria ,C
DBMS is: Cobol Marketing Director
OPSys is: VMS Configuration Data Servic
$ Range: Base $200K+ 575 Anton Blvd
Partner: Costa Mesa CA 92626-
Mfr: CDS Mfg Phone:(714)546-1892
Distributor:(714)546-1892x

Customer was Northrup Electronics Div. Includes Doc, CM, QA, Product Support, Eng Mgmt
New sys is On Line, utilizes proprietary relational structures. Has As Designed/As Planned/As Built

ECN **Engineering Change Notice**

Hdwe is: PC
DBMS is: VisBasic
OPSys is: WinDos
$ Range: $3K / Module
Partner:
Mfr: Sunbend

Bogenschut,Greg ,
President
Sunbend Corp.
11345 Highway 7
Minnetonka MN 55343-6969
Mfg Phone:(
Distributor:(612)939-9963x

Modules include Tracker, Verifier, Drawing Manager & Configuration Management
Tracker for Pending ECR's; Verifier for approved ECO's; Dwg Mgr displays CAD dwgs & CM w/BOM & CI

EDCD **Electronic Document Control Department**

Hdwe is:
DBMS is: FoxProW
OPSys is: Win 3.1
$ Range: $30K
Partner:
Mfr:

Wright ,Brad ,
Document Control Systems*
Provo UT
Mfg Phone:(

Uses a heirarchical tree to portray relationships between documents.
Developer in Salt Lake City, Heard about from student UWM#2 Irvine'93

PASS **Product Assurance Support System**

Hdwe is:
DBMS is:
OPSys is:
$ Range:
Partner:
Mfr: Compass

Swain ,Mike ,
VP Bus & Tech Operations
Compass Corp.
1945 Old Gallows Road
Vienna VA 22182-
Mfg Phone:(
Distributor:(703)556-6170x

SWCM system used as an interium protype while Info Mgmt strategy and objectives are developed.
Is an Interface & application shell arrount Softool's CCC.

PES/400 Product Engineering Support/400

Hdwe is:	MainF	
DBMS is:		
OPSys is:	AS400	IBM Corp.
$ Range:		
Partner:		
Mfr:	IBM	Mfg Phone:(

Permits maintenance of eng change data, items, BOM's & references to objects in the information
Can be integrated with MAPICS/DB database.

PhotoMemory PhotoMemory

Hdwe is:	PC	Zucker ,Mark ,
DBMS is:		Account Manager, CPIM
OPSys is:	WinMac	North Mountain Software*
$ Range:	$1750 1 User	1190 So. Bascom Ave.
Partner:		San Jose CA 95128-2026
Mfr:	NoMtnSW	Mfg Phone:(
		Distributor:(408)297-5383x

Chg Mgmt, Routing, Doc Catalogue, Doc Launching, Bar Code, Automatic Cross-Indexing
Available for Windows & Macintosh w/GUI intfc, runs across most networks. Demo Disk

SCM Software Configuration Management

Hdwe is:		Natzic ,Walter ,
DBMS is:		President & CEO
OPSys is:		Arron-Ross Corp.
$ Range:		1132 Indian Springs Dr.
Partner:	Tandem	Glendora CA 91740-
Mfr:	ARRON/ROSS	Mfg Phone:(818)963-4119
		Distributor:(818)963-4119x

Includes WBS, Doc Mgmt, BOS, Rev Status, SW Vendor Mgmt

TCC **Total Change Control**

Hdwe is:	MainF	Gill ,Abraham,
DBMS is:		
OPSys is:	DEC	CIMware Technologies Inc.
$ Range:		3031 E. LaJolla Street
Partner:		Anaheim CA 92806-1303
Mfr:	CIMware	Mfg Phone:(
		Distributor:(415)962-1700xGill

This might be what was CMsoft from Telavie Israel.

TEDS **Technical Data Configuration Management System**

Hdwe is:	Multi	
DBMS is:	See Note	
OPSys is:		VSE Corp.
$ Range:		2760 Eisenhower Ave.
Partner:		Alexandria VA 22314-4587
Mfr:	VSE	Mfg Phone:(
		Distributor:(703)329-2633x

Targeted at prime contractors or govmt agencies. Has links to DoD Logistics databases.
Available in Oracle, Informix & IDMS-R. Army Std CM System.

Compound Doc

InForms **Informs**

Hdwe is:	Multi	
DBMS is:		
OPSys is:	Multi	Word-Perfect Corp.
$ Range:		1555 N. Technology Way
Partner:		Orem UT 84057-
Mfr:	WordPerf	Mfg Phone:(
	1-(800)451-5151	Distributor:(801)225-5000x

Windows style electronic forms builder w/database & e-mail interfaces on Windows DOS, MAC, OS2,
No Graphics? ASCII, dBase, FoxPro, Paradox, Informix, SQL Server, Oracle, SQL Base, Sybase &

Linkage **Automated Manufacturing Information Flow**

Hdwe is:	WS	
DBMS is:		
OPSys is:	UNIX	Cimlinc, Inc. **
$ Range:	$20K+	
Partner:		
Mfr:	CIMLINC	Mfg Phone:(708)250-0090
	1-(800)225-7943	

Interface to Oracle, Ingres, Sybase, Empress, SQL, ASCII, IGES, ...

Sequel

Hdwe is:		Miller ,Phoebe ,
DBMS is:		Distributor
OPSys is:		Techgnosis, Inc.
$ Range:		
Partner:		
Mfr:		Mfg Phone:(

Like Linkage

Tools

CES **AMAPS CES, Contract Engineering System**

Hdwe is:	MainF	Guiess ,Bob ,
DBMS is:	CADlinc	
OPSys is:	IBM	Mgmt Sciences America
$ Range:		3445 Peachtree Road, N.E.
Partner:	Tandem	Atlanta GA 30326--127
Mfr:	MSA	Mfg Phone:(
		Distributor:(404)239-2400x

Moves CAD attributes to the PC.
Partners with Boeing, EDS. Factory Control Mgmt Sys runs on Tandem. Old Comserv Corp.

DFT-PAC **Draft_Pac Bill of Material Database Generator**

Hdwe is:	PC	Evans ,Michael,
DBMS is:	DBASE	National Sales
OPSys is:	MSDOS	Cadkey, Inc.
$ Range:		4 Griffin Road North
Partner:		Windsor CT 06095-
Mfr:	Cadkey	Mfg Phone:(
	1-(800)394-2231	Distributor:(203)298-8888x

Works with CADKEY to integrate a BOM database with the CAD package.
No external database is required, can be used with any existing DBASE compatible database.

DS **Documentation System**

Hdwe is:	PC	Stephens ,Bert ,
DBMS is:		President
OPSys is:		Stephens Engineering
$ Range:	$109.00	
Partner:		San Franscis CA
Mfr:	Stephens	Mfg Phone:(408)730-2240
		Distributor:(408)730-2240x

A 219 category P/N sys for inventory, dwgs, assys, PWBs, Proc, specs & main models.
Preprinted forms include ECO, ECR, & MDR. This is a manual & forms, NOT AUTOMATED!

Computer Aided Design/Engineering

CAD/CAE

ANVIL 5000 Anvil 5000

Hdwe is: Multi — Devere ,Gerald ,J
DBMS is: — VP
OPSys is: — Mfg & Consulting Services
$ Range:
Partner: — Irvine CA
Mfr: MCS — Mfg Phone:(

AutoCAD AutoCAD & AutoCAD Designer

Hdwe is: PC
DBMS is:
OPSys is: DOS — Autodesk, Inc.
$ Range:
Partner: — CA
Mfr: AutoDesk — Mfg Phone:(
Distributor:(415)491-8223x

Developed primarily for architectural apps. Designer is parametric, feature-based solid modeler.
SQL access added in ACAD Data Extension (ADE) module 4Q93

CADKEY CADKEY

Hdwe is: — Evans ,Michael,
DBMS is: — National Sales
OPSys is: DOS — Cadkey, Inc.
$ Range: — 4 Griffin Road North
Partner: — Windsor CT 06095-
Mfr: Cadkey — Mfg Phone:(
1-(800)654-3413 — Distributor:(203)298-8888x

Also offer ANALYSIS and CUTTING EDGE

CADNETICS Cadnetics

Hdwe is:
DBMS is:
OPSys is: — Intergraph
$ Range:
Partner:
Mfr: — Mfg Phone:(

Printed Circuit Board (PCB) system

CATIA **CATIA Version 3**

Hdwe is: MainF
DBMS is:
OPSys is: IBM MVS IBM Corp.
$ Range:
Partner:
Mfr: IBM Mfg Phone:(

w/3D Design; Building Design; Advanced Surfaces; Interface; NC; Solids Geo; Kinematics; Robotics;
& w/Drafting; Library & Image Design

CIM CAD **CIM CAD 2 1/2 or 3D**

Hdwe is: WS Tolvstad ,Eric ,
DBMS is: UNIX Account Coordinator
OPSys is: Great River Sys*
$ Range: 4252 Park Ave
Partner: Mpls MN 55407-
Mfr: CIMLINC Mfg Phone:(708)250-0090
1-(800)225-7943 Distributor:(612)686-0995x 10

DDN **Design, Drafting and Numeric Control**

Hdwe is: WS
DBMS is:
OPSys is: ICEM Systems Inc.
$ Range:
Partner:
Mfr: ICEM Sys Mfg Phone:(612)853-8100

Generic CAD Generic CAD

Hdwe is: PC
DBMS is:
OPSys is: DOS Autodesk, Inc.
$ Range:
Partner: CA
Mfr: AutoDesk Mfg Phone:(
Distributor:(415)491-8223x

Attributes such as PL, can be output in ASCII or Lotus 123 format

I-DEAS

Hdwe is:		
DBMS is:	Oracle	
OPSys is:		SDRC *
$ Range:	$25 - 100K	
Partner:		
Mfr:	SDRC 1-(800)922-7372	Mfg Phone:(714)542-2201

MasterSeries Master Series

Hdwe is:		
DBMS is:		
OPSys is:		SDRC *
$ Range:		
Partner:		
Mfr:	SDRC 1-(800)922-7372	Mfg Phone:(714)542-2201

ME10 2D CAD

Hdwe is:	WS	
DBMS is:		
OPSys is:	HP-UX	Hewlett-Packard Co.
$ Range:		
Partner:		Palo Alto CA 94303-0890
Mfr:	HP	Mfg Phone:(Distributor:(415)857-1501x

MicroStation Microstation

Hdwe is:		
DBMS is:		
OPSys is:		Intergraph
$ Range:		
Partner:		
Mfr:	Bentley	Mfg Phone:(

Intergraphs flagship CAD line. Bentley Systems is the developer
Bentley Systems, Exton PA, will take over the line in '95

Patran 3	**Patran Analysis Software**	
Hdwe is:		
DBMS is:		
OPSys is:		PDA Engineering
$ Range:		
Partner:		
Mfr:	PDA	Mfg Phone:(708)882-2223

PD CAD	**Personal Designer CAD**	
Hdwe is:		Reda ,Tracey ,
DBMS is:		Account Representative
OPSys is:		Computervision Corp.*
$ Range:		100 Crosby Drive
Partner:		Bedford MA 01730-1480
Mfr:	CV	Mfg Phone:(
	1-(800)786-2231	Distributor:(800)RUN-CAD1x

Has CADDS and DXF translators

PDGS	**Product Design Graphic System**	
Hdwe is:		Reda ,Tracey ,
DBMS is:		Account Representative
OPSys is:		Computervision Corp.*
$ Range:		100 Crosby Drive
Partner:		Bedford MA 01730-1480
Mfr:	CV	Mfg Phone:(
	1-(800)786-2231	Distributor:(800)RUN-CAD1x

ProEng	**Pro/ENGINEER**	
Hdwe is:	Multi	Silver ,Robin ,
DBMS is:		Product Manager
OPSys is:	Unix	Parametric Technology
$ Range:	$17K +	128 Technology Drive
Partner:		Waltham MA 02154-
Mfr:	ParaMetric	Mfg Phone:(
		Distributor:(617)894-7111x

Powerful modeler with associativitity, not strong in 2D CAD

Unigraphics **Unigraphics**

Hdwe is: WS
DBMS is:
OPSys is: EDS*
$ Range: 13736 Riverport Drive
Partner: Maryland Hts MO 63043-
Mfr: EDS Mfg Phone:(
Distributor:(800)344-5900x

Recently (Fall '91) EDS purchased from McDonnell Douglas.

Source Code

Razor **Problem Tracking System**

Hdwe is: WS Ivory ,John ,
DBMS is: President
OPSys is: Sun OS Tower Concepts, Inc.
$ Range: $395 103 Sylvan Way
Partner: New Hartford NY 13413-
Mfr: Tower Mfg Phone:(
Distributor:(315)724-3540x

Front end tool, works with SCCS & RCS. Problem Tracking, File Version Control, Release, e-mail

Users can add functionality via unix scripts, & attributes to Razor screens & windows

Tools

CADoverlay **CADoverlay**

Hdwe is: Althaus ,Don ,
DBMS is: Sales
OPSys is: Engineering Reprographics
$ Range:
Partner: Plymouth MN
Mfr: ImageSys Mfg Phone:(
Distributor:(612)473-2902x

HiJack **HiJack Pro Image Converter and Paintbrush**

Hdwe is:		
DBMS is:		
OPSys is:	Win 3.0	Inset Systems
$ Range:	$249	71 Commerce Drive
Partner:		Brookfield CT 06804-
Mfr:	HiJack	Mfg Phone:(203)740-2400
		Distributor:(203)740-2400x

CGM Raster converter, SGML to TIFF, Postscript to TIFF. V2+ has reverse engineering.
Graphics Utility, w/view, enhance, convert & print in DOS or Windows, Sun & Mac

IGESleaf **IGES to Interleaf**

Hdwe is:	Multi	Hess ,Charlen,A
DBMS is:	N/A	Director of Marketing
OPSys is:	DosUnix	IGES Data Analysis
$ Range:		2001 N. Janice Ave
Partner:		Melrose Park IL 60106-
Mfr:	IDA	Mfg Phone:(708)344-1815
		Distributor:(708)344-1815x334

Runs on Sun 3, Sparc, DEC VAX/VMS, Apollo, RS6000 & MS-DOS

IGESMaker **IGES to FrameMaker**

Hdwe is:	Multi	Hess ,Charlen,A
DBMS is:	N/A	Director of Marketing
OPSys is:	DosUnix	IGES Data Analysis
$ Range:		2001 N. Janice Ave
Partner:		Melrose Park IL 60106-
Mfr:	IDA	Mfg Phone:(708)344-1815
		Distributor:(708)344-1815x334

Creates editable MIF file

Integrator **The Integrator**
Hdwe is: Multi
DBMS is:
OPSys is: C-TAD Systems, Inc.
$ Range: 3025 Boradwalk
Partner: Ann Arbor MI 48108-
Mfr: C-TAD Mfg Phone:(
Distributor:(313)665-3287x

Custom Translation Software.
ACAD Unigraphics IGES CADAM I-DEAS CGS STL EUCLID VDA CATIA CV PDGS STRIM TOYOTA MAZDA HONDA NISSAN

TIMSR12 **Total Information Management System (ACAD R12)**
Hdwe is: WS Galpin ,Richard,
DBMS is: President, CEO
OPSys is: UnixDos Object Graphics
$ Range: 400 Stierlin Road
Partner: MountainView CA 94043-
Mfr: ObjectGrph Mfg Phone:(
Distributor:(415)968-1500x

Linked to ACAD, Links to DBMS, Analysis Tools, Reports. ACAD Viewer available
Targeted at Geographical Information Systems. Demo

Tracer **Tracer for AutoCAD**
Hdwe is: Kehn ,Don ,E
DBMS is: VP Engineering
OPSys is: Information Graphics Sys
$ Range: 2511 55th Street
Partner: Boulder CO 80301-
Mfr: IGS Mfg Phone:(
Distributor:(303)449-1110x

Raster-to-Vector conversion tool for AutoCAD, w/Raster snap & overlay, Interactive line tracing.
Targeted at Geographical Information Systems. Based on CADCore/ Tracer.

Product Data Management (PDM)

File Control

10CAD 10CAD Eng Data Mgmt System

Hdwe is: Hayes ,Dale ,
DBMS is:
OPSys is: ACS Telecom, Inc.
$ Range: $10,000+ 25825 Eshelman Ave.
Partner: Lomita CA 90717-3220
Mfr: ACSTelecom Mfg Phone:(

File Control supports ACAD, WordPerfect, MultiMate
Supports "BTrieve" Format, "Novell & Ethernet Network OPS"

AM Workflow AutoManager Document Management for AutoCAD

Hdwe is: Sneassert ,Philip ,
DBMS is:
OPSys is: Cyco International*
$ Range: $195 - $3895 1908 Cliff Valley Way
Partner: Atlanta GA 30329-
Mfr: Cyco Mfg Phone:(
1-(800)323-2926 Distributor:(404)634-3302x112

File Control supports ACAD, possibly customizable,
AM-WorkFlow, AutoManager Classic & Professional; 3 Demo Disks

AR-MARC Multi-Media Data Storage & Archive

Hdwe is: Natzic ,Walter ,
DBMS is: President & CEO
OPSys is: Arron-Ross Corp.
$ Range: 1132 Indian Springs Dr.
Partner: Tandem Glendora CA 91740-
Mfr: ARRON/ROSS Mfg Phone:(818)963-4119
Distributor:(818)963-4119x

Include tape and optical archive

AutoEDMS **AutoEDMS**
Hdwe is: PC Hayes ,Dale ,
DBMS is:
OPSys is: Novell ACS Telecom, Inc.
$ Range: $10,000+ 25825 Eshelman Ave.
Partner: Lomita CA 90717-3220
Mfr: ACS Mfg Phone:(
Supports AutoCAD, WordPerfect & Multimate.

BOSS **Document Manager**
Hdwe is: Multi Williams ,Bob ,
DBMS is: Sybase VP Marketing
OPSys is: UnixWin BOSS Logic
$ Range: $500 & Up
Partner: MountainView CA
Mfr: BOSS Mfg Phone:(
Distributor:(415)903-7000x
Lets users set up routing of the doc and establish which users can access the doc throughout
Being ported to Windows 3.x, Unix, Next

CADEXnet **CADEXnet**
Hdwe is: PC Cann ,David ,
DBMS is:
OPSys is: DOS Database Applications Inc
$ Range: $695 to $60K 14 Admiralty Place
Partner: Redwood City CA 94065-
Mfr: DBA Mfg Phone:(
Distributor:(415)593-3477x
Works with AutoCAD Uses an Index Card metaphore.
Can run on SUN (UNIX) with PC-NFS or on VAX VMS. Windows in '92,

Cadmandu **Cadmandu - CAD Management & Drawing Utilities**
Hdwe is: Multi Millan ,Bob ,J
DBMS is: C-Tree Sales Manager
OPSys is: UNIXdos BaraTek, Inc.
$ Range: $1K to $1.5K 1550 N. Northwest Hwy.
Partner: Park Ridge IL 60068-
Mfr: Baratek Mfg Phone:(
Distributor:(708)803-6363x
CAD management with Redline & View
Unix, DOS & Windows versions avail. Supports CADKey, AutoCAD. Demo Disk

CDM/CDMA CATIA DataMgmt CATIA DataMgmt Access

Hdwe is:	MainF	
DBMS is:	DB2 &	
OPSys is:	MVS- VM	IBM
$ Range:		3200 Windy Hill Road
Partner:		Marietta GA 30067-
Mfr:	IBM	Mfg Phone:(
		Distributor:(404)835-8522x

No Change Mgmt or Program Mgmt. Limited functionality
One of few for IBM, interface plans with ProductManager currently unclear. SQL/DS supported.

CMS Configuration Management System

Hdwe is:	WS	Carney ,Jim ,
DBMS is:	Oracle	President
OPSys is:	Sun 4.1	Workgroup Technologies Co
$ Range:	$24,000+	140 Second Ave.
Partner:		Waltham MA 02254-
Mfr:	Wkgrptech	Mfg Phone:(
		Distributor:(617)895-1500x3017

CM system available, but not mature yet. GUI interface, good file control.
Entered the file control market in 1986

DCM CADnet Document Control Manager

Hdwe is:	WS	Sullivan ,Pat ,
DBMS is:		
OPSys is:	Unix	Computer Signal Corp.
$ Range:	$5,995	2420 Camino Ramon
Partner:		San Ramon CA 94583-
Mfr:	CSC	Mfg Phone:(415)275-0990
		Distributor:(415)275-0990x

Document Library; user, directory, permission & backup mgmt; Check in/out; & print/plot mgr
Works with AutoCAD on Unix & MS-DOS over TCP/IP & FTP networks.

DDMS Document, Drawing & Image Management System

Hdwe is:		Natzic ,Walter ,
DBMS is:		President & CEO
OPSys is:		Arron-Ross Corp.
$ Range:		1132 Indian Springs Dr.
Partner:	Tandem	Glendora CA 91740-
Mfr:	ARRON/ROSS	Mfg Phone:(818)963-4119

Distributor:(818)963-4119x

Includes history log of Revs, Doc/Dwg/Image attributes, & On-line optical storage

I saw demo at Ref site in Milwaukee. Is a highly tailered implementation. This one ran on Tandem.

DEED **DEED**

Hdwe is:	Multi	BKS ,Lee ,
DBMS is:		Marketing
OPSys is:	Sun Dos	Bionic Knight Software
$ Range:	PC $995 & up	
Partner:		Raleigh NC 27624-
Mfr:	BKS	Mfg Phone:(
		Distributor:(919)847-1531x

Contains BOM data, Drawing Mgmt and view, has a DBMS and CAD interface. Can interface to MRP

Utilizes the "C" language, available for PC or SUN networks. $2500 on Unix; Modular; Demo Disk

DMCS **Data Management and Control System**

Hdwe is:	MainF	Shoaf ,Steve ,
DBMS is:	Oracle	Mgr Tech Support
OPSys is:		SDRC
$ Range:	$25 - 100K	2000 Eastman Dr
Partner:		Milford OH 45150-2789
Mfr:	MetaPhase	Mfg Phone:(714)542-2201
	1-(800)576-2400	Distributor:(513)576-2400x

DMCS will be the Metaphase V2.0 product in early 1994. That is an integration of DMCS & EDL.

DMS

Hdwe is:		
DBMS is:		
OPSys is:		Vanderoot
$ Range:		
Partner:		
Mfr:	Vanderoot	Mfg Phone:(

DMX **Data Management Exchange**

Hdwe is:		Natzic ,Walter ,
DBMS is:		President & CEO
OPSys is:		Arron-Ross Corp.
$ Range:		1132 Indian Springs Dr.
Partner:	Tandem	Glendora CA 91740-
Mfr:	ARRON/ROSS	Mfg Phone:(818)963-4119
		Distributor:(818)963-4119x

They are planning to make products 'database independent'.

DocMgr **Document Manager**

Hdwe is:		Pray ,Chris ,
DBMS is:	Oracle	Marketing
OPSys is:		Cimage Corp.
$ Range:		3885 Research Park Dr
Partner:		Ann Arbor MI 48108-
Mfr:	Cimage	Mfg Phone:(
		Distributor:(313)761-6550x

DwgLibrarian **Drawing Librarian Professional (DP)**

Hdwe is:		
DBMS is:		
OPSys is:		SoftSource
$ Range:	$500	
Partner:		
Mfr:	SoftSource	Mfg Phone:(206)676-0999

New features; red-lining, included script language, customizable user interface.

DwgMgmt **Drawing Management**

Hdwe is:	WS	
DBMS is:		
OPSys is:	UNIX	Cimlinc, Inc.
$ Range:		
Partner:		
Mfr:	CIMLINC	Mfg Phone:(708)250-0090
	1-(800)225-7943	

EDCS II **Engineering Data Control System II**

Hdwe is:	MainF	
DBMS is:	Rdb	
OPSys is:	VAX	Digital Equipment Corp.

$ Range:	$12 - 70K	200 Forest Street
Partner:		Marlboro MA 01752-
Mfr:	DEC	Mfg Phone:(617)264-1111

Not actively pursuing new implementations.

EDL **Engineering Data Library**

Hdwe is:	Multi	
DBMS is:		
OPSys is:	UNIX	Control Data Sys, Inc.
$ Range:	$26.5K/seat	
Partner:		
Mfr:	MetaPhase	Mfg Phone:(612)853-6637

Runs Memory Data Manager in Unix environment, Oracle supported, BASIS+ on Mainframe.
Leader in Eng Data Management systems since 1982. Working with CMstat.

EDM **Engineering Data Management**

Hdwe is:	MainF	Reda ,Tracey ,
DBMS is:	Oracle	Account Representative
OPSys is:	Unix	Computervision Corp.
$ Range:	$75 - 150K	100 Crosby Drive
Partner:		Bedford MA 01730-1480
Mfr:	CV	Mfg Phone:(
	1-(800)786-2231	Distributor:(800)RUN-CAD1x

Powerful, 1 of top 4 or 5 PDM systems available.
SQL/DS (IBM VM/SP & VM/HPO); Oracle (SUN & DEC VAX/VMS), Prime 50

EXPress **EXPress**

Hdwe is:		Borja ,Phil ,
DBMS is:	RDBMS's	VP Marketing
OPSys is:		EXP Group, Inc.
$ Range:		15042 Espola Road
Partner:		Poway CA 92064-
Mfr:	EXP	Mfg Phone:(

Object Oriented approach on Sybase, Oracle, Ingres
Has EXPress Chg Order, Proj Mgmt, View/Redline, Image Mgmt, etc.

FAIM **Cimplex-Factory Automation Information Management**

Hdwe is:	MainF	Ezzard ,Ralph ,
DBMS is:	Propriet	Consulting Director
OPSys is:	IBM	Automation Tech Products
$ Range:	$100 - 250K	
Partner:		Campbell CA
Mfr:	ATP	Mfg Phone:(408)370-4000

IBM Hardware

FileMgr **Synergis Network FileManager**

Hdwe is:	WS	Lamond ,Scott ,
DBMS is:	db3 like	SW Marketing Coordinator
OPSys is:	Sun/Dos	Synergis Technologies Inc
$ Range:	$1,995 & up	593 Skippack Pike
Partner:		Blue Bell PA 19422-
Mfr:	Synergis	Mfg Phone:(
	1-(800)836-5440	Distributor:(215)643-9050x

Integrated with ACAD. Has File control, releaser, viewer and launcher. Database is proprietary, but compatable with dBase 3. Demo Disk

I/MAN **INFOMANAGER Version 2**

Hdwe is:	WS	
DBMS is:	Oracle	
OPSys is:	VAX/UNI	EDS
$ Range:		13736 Riverport Drive
Partner:		Maryland Hts MO 63043-
Mfr:	EDS	Mfg Phone:(
		Distributor:(800)344-5900x

Based on student comment in San Diego, "a crippled release of a system that may have good

Was McDonnell Douglas. Product based on a good (BOM), Ingres in '92.

I/PDM **Intergraph Product Data Management**

Hdwe is:	Multi	Nolte ,Kurt ,A
DBMS is:	Informix	Sales Representative
OPSys is:	UNIX	Intergraph Corp.
$ Range:	$2 - 90K	20725 Watertown Road
Partner:		Waukesha WI 53186-
Mfr:	Intergraph	Mfg Phone:(414)798-1495
		Distributor:(414)798-1495x

VMS (DEC), Unix (WS)

Used for Parts Lists and BOM, has interfaces to various relational databases

Imaxis Imaxis

Hdwe is: MacIn

DBMS is:

OPSys is: Systems Engr'g Solutions

$ Range: $2695 4 user

Partner: Dunn Loring VA

Mfr: SES Mfg Phone:(

Document Management System for Macintosh and workgroups
Currently in Beta testing, due to ship Jan '94

Intellect Intellectual Property

Hdwe is: Meyer ,Jim ,

DBMS is: Oracle VP Doc Mgmt Group

OPSys is: Unix Interleaf, Inc.

$ Range: $130K 40/usr 9 Hillside Ave

Partner: Waltham MA 02154-

Mfr: Interleaf Mfg Phone:(

1-(800)223-6638 Distributor:(617)290-0710x

Packaged with the Relational Document Manager (RDM) product.
For accessing intectual property such as marketing material or engr'g information.

NovaManage NovaManage CAD management

Hdwe is: WS Dunham ,Scott ,

DBMS is: Oracle

OPSys is: Cli/Svr Novasoft Systems, Inc.

$ Range: $75K 8 users 8 New England Exec Park

Partner: Burlington MA 01803-

Mfr: Novasoft Mfg Phone:(

Distributor:(617)272-0300x

Allows users to manage, track & control a variety of documents.
Supports a variety of database management packages. Adindu designed this one

OPEN/ OPEN/Profound and OPEN/Workflow

Hdwe is: WS Ryan ,Burce ,
DBMS is: VP & General Manager
OPSys is: AIX Wang Laboratories, Inc.
$ Range: 1 Industrial Ave.
Partner: Lowell MA 01851-
Mfr: Wang Mfg Phone:(

Running on IBM RS/6000

PE/WorkMgr PE/Work Manager

Hdwe is: WS
DBMS is: Oracle
OPSys is: HP-UX Hewlett-Packard Co.
$ Range: $10K +
Partner: Palo Alto CA 94303-0890
Mfr: HP Mfg Phone:(
Distributor:(415)857-1501x

Solid PDM product for HP CAD/CAE environments, has optional BOM and GT modules.

Product is small, simple, easy to use. Lacks robust rel or change modules.

PIMS Sherpa Design Management System

Hdwe is: MainF McFadden ,Tim ,
DBMS is: Propriet
OPSys is: Sherpa Corp.
$ Range: $30K and up 13074 Soaring Bird Point
Partner: Poway CA 92128-
Mfr: Sherpa Mfg Phone:(408)433-0455
Distributor:(619)486-0152x

The Ingres DBMS is behind their proprietary 'C' interface, but the data doesn't appear accessible

ProPDM Pro/PDM

Hdwe is: Silver ,Robin ,
DBMS is: Product Manager
OPSys is: Parametric Technology
$ Range: $5,000/seat 128 Technology Drive
Partner: Waltham MA 02154-
Mfr: ParaMetric Mfg Phone:(
Distributor:(617)894-7111x

RDM Relational Document Manager

Hdwe is: Hanzlik ,Vern ,

DBMS is:	Oracle	Sales
OPSys is:	Unix	Tech Pub Solutions, Inc.
$ Range:	$40K/8 seats	5500 Lincoln Drive
Partner:		Minneapolis MN 55436-
Mfr:	Interleaf	Mfg Phone:(
	1-(800)223-6638	Distributor:(612)938-4490x

Has Vault, Doc routings & msgs, Doc CM (manages doc interelationships) & Workflow Mgmt.
May be LISP based.

RMS **Release Management System**

Hdwe is:		Natzic ,Walter ,
DBMS is:		President & CEO
OPSys is:		Arron-Ross Corp.
$ Range:		1132 Indian Springs Dr.
Partner:	Tandem	Glendora CA 91740-
Mfr:	ARRON/ROSS	Mfg Phone:(818)963-4119
		Distributor:(818)963-4119x

Includes document WIP, document routing, date/time stamping and Formal Release

TDM **Technical Data Management**

Hdwe is:	PC	
DBMS is:		
OPSys is:	DOS	Autodesk, Inc.
$ Range:		
Partner:		CA
Mfr:	AutoDesk	Mfg Phone:(
		Distributor:(415)491-8223x

Target marketing date is 1Q95

TDMS **Technical Document Management System**

Hdwe is:	Multi	Clark ,Blaine ,
DBMS is:	Oracle	VP of Sales
OPSys is:	VAX/VMS	Access Corp.
$ Range:	$.3M $1M	1011 Glendale-Milford Rd
Partner:		Cincinnati OH 45215-
Mfr:	ACCESS	Mfg Phone:(
		Distributor:(513)782-8300x

Key element is EDICS.
Rel-Y; EO-Y; BOM-N; GT-C; PM-N. Supports CALS. May be on an OODBMS.

VAULT **Adra Vault**

Hdwe is: Multi — Adra ,1 ,
DBMS is: Empress
OPSys is: VAX/VMS — Adra Systems, Inc.
$ Range: $12K $65K — 59 Technology Drive
Partner: — Lowell MA 01851-
Mfr: ADRA — Mfg Phone:(
Distributor:(508)937-3700x

Marketed primarily as support to Adras CADRA-III CAD/CAM systems. Rel-Y; EO-Y; BOM-Y; GT-N; PM-N.

VisionAel **VisionAel, Engineering Management System**

Hdwe is:
DBMS is:
OPSys is: — Advanced Graphix
$ Range:
Partner:
Mfr: AdvGraphic — Mfg Phone:(

Comp in Oklahoma, Info from Mark Teetsel

Image Sys

1stSCAN **FirstScan**

Hdwe is: WS
DBMS is:
OPSys is: Unix — Aurora Technologies, Inc.
$ Range: $995
Partner:
Mfr: Aurora — Mfg Phone:(617)290-4800

Formats; TIFF, SunRaster, PostScript. Requires your HP ScanJet Scanner.

DocuData **DocuData - Client/Server document mgmt application**

Hdwe is:
DBMS is:
OPSys is: — LaserData, Inc.
$ Range: $2,000 /seat
Partner: — Tyngsboro MA
Mfr: LaserData — Mfg Phone:(

DocuData includes a viewer & database browser.
DocuFlow - a workflow module that works with Action Workflow System by Action Technologies.

DocuPlex DocuPlex

Hdwe is:	WS	Banie ,James ,
DBMS is:	Oracle ?	Sr Systems Analyst
OPSys is:	Unix/PC	Xerox Corp.
$ Range:		2301 West 22nd Street
Partner:		Oak Brook IL 60521-
Mfr:	Xerox	Mfg Phone:(
		Distributor:(708)572-9585x

Prelim Info: has controlled distribution, uses "Tree Structure" to file docs & define configuration
A prospect saw demo at InfoMart, hard to tell if product has any depth.
VERY NEW

EasyScan EasyScan Image Scanning System

Hdwe is:	WS	
DBMS is:		
OPSys is:	Unix	Pectronics Corp.
$ Range:	$2500+	
Partner:		
Mfr:	Pectronics	Mfg Phone:(408)867-3180

Formats; TIFF, SunRaster, PostScript or EPS.
Also offer EasyRead OCR @ $995. DigitalPhoto Image Retouching SW.

EDICS Engineering Document Image Management System

Hdwe is:	MainF	Clark ,Blaine ,
DBMS is:	Oracle	VP of Sales
OPSys is:	AS400	Access Corp.
$ Range:		1011 Glendale-Milford Rd
Partner:		Cincinnati OH 45215-
Mfr:	ACCESS	Mfg Phone:(
		Distributor:(513)782-8300x

EDICS/ECC (change control) & /RM (release mgr)
Also have logging of files on Unify PC DB for a form of config control

EIPM **Electronic Information Process Management**

Hdwe is: Mayo ,John ,
DBMS is:
OPSys is: Computer Task Group, Inc.
$ Range: 7918 Jones Branch Dr #500
Partner: McLean VA 22102-9613
Mfr: CTG Mfg Phone:(
Distributor:(703)790-1557x

FileMagicVis **File Magic Vision**

Hdwe is: Graham ,Michael,
DBMS is: Sr. VP Marketing
OPSys is: Westbrook Technologies
$ Range: $199
Partner: Westbrook CT
Mfr: Westbrook Mfg Phone:(

features include a database, doc viewer, and scanner integration. Supports 200+ formats
including video capture and photographs.

FORMPRO **Aaron-Ross Forms Processing**

Hdwe is: Natzic ,Walter ,
DBMS is: President & CEO
OPSys is: Arron-Ross Corp.
$ Range: 1132 Indian Springs Dr.
Partner: Tandem Glendora CA 91740-
Mfr: ARRON/ROSS Mfg Phone:(818)963-4119
Distributor:(818)963-4119x

Includes Scanned, FAXed, & WS images in file folder control

HyperviewRed **Hyperview Redline**

Hdwe is: WS
DBMS is:
OPSys is: Unix Techview Corp.
$ Range: $695
Partner:
Mfr: TechView Mfg Phone:(

Drawing, viewing & redlining for Sun SPARC. Handles Raster & Vector

Image Plus **ImagePlus File Cabinet**

Hdwe is: AS400
DBMS is:

OPSys is: IBM
$ Range: 3200 Windy Hill Road
Partner: Marietta GA 30067-
Mfr: IBM Mfg Phone:(
Distributor:(404)835-8522x

1st of IBM Server Series Solutions, includes IBM standalone optical drive

ImageFlow Plexus ImageFlow & Work Process Automation

Hdwe is: Dunn ,Gary ,
DBMS is:
OPSys is: Recognition International
$ Range: 1310 Chesapeake Terrace
Partner: Sunnyvale CA 94089-
Mfr: Recognitio Mfg Phone:(
Distributor:(408)743-4300x

ImageMaster ImageMaster

Hdwe is: Kelly ,Robert ,P
DBMS is: Marketing Executive
OPSys is: SeeNote Cimage Corp.
$ Range: 3885 Research Park Dr
Partner: Ann Arbor MI 48108-
Mfr: Cimage Mfg Phone:(
Distributor:(313)761-6550x

Raster image scanning, QA, view, & redline capabilities.
OP Sys - DOS, PS/2, RS/6000 w AIX, & Unix (Suns) or DEC ULTRIX.

Imaging Ultimation Imaging 1.1

Hdwe is:
DBMS is:
OPSys is: Ultimation Inc.
$ Range: $8K 5 users
Partner: Herndon VA
Mfr: UI Mfg Phone:(

Lets users store, retrieve, print & export color, gray scale & black/white docs.
Users drag and drop icons to perform most functions. Can store folders within folders.

InfoImage **Info Image Folder**
Hdwe is:
DBMS is:
OPSys is: Unisys
$ Range:
Partner:
Mfr: Unisys Mfg Phone:(

IntelliVue **IntelliVue**
Hdwe is:
DBMS is:
OPSys is: Intellinetics
$ Range:
Partner:
Mfr: Intellinet Mfg Phone:(

First imaging system built solely on CD-ROM instead of WORM. Lets users store 15 - 20,000 documents on 1 CD-Recordable. Avoids expensive Jukeboxes.

MEDIS **Modular Electronic Document Information Solution**
Hdwe is: WS
DBMS is: Oracle
OPSys is: Unix InterLinear Technology
$ Range: 1320 Harbor Bay Parkway
Partner: Alameda CA 94501-
Mfr: ILT Mfg Phone:(
Distributor:(510)748-6850x

Myriad **Myriad**
Hdwe is: Heath ,Gary ,
DBMS is:
OPSys is: Informative Graphics
$ Range: $395
Partner: Phoenix AZ
Mfr: InfoGraphi Mfg Phone:(

Now tied with SHERPd per Richard Gross

Open/Image **Open/Image**
Hdwe is: Ryan ,Burce ,
DBMS is: VP & General Manager
OPSys is: UnixWin Wang Laboratories, Inc.

$ Range:	$495 /user	1 Industrial Ave.
Partner:		Lowell MA 01851-
Mfr:	Wang	Mfg Phone:(

Running on IBM RS/6000. May be database independant

PixTex/EFS PixTex/EFS (Electronic File System)

Hdwe is:	MainF	
DBMS is:		
OPSys is:	VAX/VMS	Excalibur Technologies
$ Range:		9255 Towne Centre Dr.
Partner:		San Diego CA 92121-
Mfr:	Excalibur	Mfg Phone:(

Avnet Cmptr Inc. will package PixTex/EFS, DEC hdwe & Calera Recongition Sys scanners for turnkey

PowerImage PowerImage

Hdwe is:	Multi	Strasnick ,Barry ,
DBMS is:		Founder
OPSys is:	UnixDos	Desktop Advantage
$ Range:		796 Beacon Street
Partner:		Newton MA 02159-
Mfr:	Desktop	Mfg Phone:(617)965-6590
		Distributor:(617)965-6590x

See Datasheet for details

Replica Replica for Windows [& Replica Viewer]

Hdwe is:	PC	Marketing ,FCI ,
DBMS is:	N/A	Marketing
OPSys is:	Win	Farallon Computing, Inc.
$ Range:	$99	2470 Mariner Square Loop
Partner:		Alameda CA 94501-1010
Mfr:	Farallon	Mfg Phone:(
		Distributor:(510)814-5000x

Distribute windows WYSIWYG docs, they will see a copy without the orig application.

Retains all fonts, formatting & graphics. Mac version comming. Demo Disk Avail

Sherpa/View **Sherpa/View**

Hdwe is:	MainF	McFadden ,Tim ,
DBMS is:	Propriet	
OPSys is:		Sherpa Corp.
$ Range:	$745+	13074 Soaring Bird Point
Partner:		Poway CA 92128-
Mfr:	Sherpa	Mfg Phone:(408)433-0455
		Distributor:(619)486-0152x

Doc viewing & markup tool integrated with Sherpa/PIMS

ViewBase **ViewBase**

Hdwe is:		Althaus ,Don ,
DBMS is:		Sales
OPSys is:		Engineering Reprographics
$ Range:		
Partner:		Plymouth MN
Mfr:	ImageSys	Mfg Phone:(
		Distributor:(612)473-2902x

Route, View & Redline

ViewStar

Hdwe is:	PC	Wehrle ,Barbara,
DBMS is:		Dir of Marketing
OPSys is:	Win/Dos	Viewstar Corp.
$ Range:	$800 to$175K	
Partner:		Emeryville CA
Mfr:	Viewstar	Mfg Phone:(510)652-7827
		Distributor:(510)652-7827x

Route, View & Redline SQL supported, search date, author, keywords; Auto Archiving

OCR, Grp3&4 Fax, TIFF, CCITT supported. 8MB Ram req'd.

Redline

AutoVue **AutoVue-Redline**

Hdwe is:	PC	Verelli ,Angelo ,
DBMS is:		Sales
OPSys is:	DosUnix	Cimmetry Systems Inc.
$ Range:	$250 & up	1430 Mass Ave
Partner:		Cambridge MA 02138-3810
Mfr:	CSI	Mfg Phone:(
	1-(800)361-1904	Distributor:(514)735-3219x

Handles over 30 CAD Vector, HPGL & Raster Formats

RedLine Module, DDBMS Module (searchable document library), Integrator Module combines Redline.

CADleaf Red CADleaf Redliner 1.0

Hdwe is:	WS	
DBMS is:		
OPSys is:	Unix	Carberry Technology, Inc.
$ Range:		600 Suffolk Street
Partner:		Lowell MA 01854-USA
Mfr:	CTI	Mfg Phone:(
		Distributor:(508)970-5358x

Viewing & redlining graphical data. IGES, DXF, HP-GL(2), CalComp 906/907 & CGM vectors.
CCITT gr4 & TIFF rasters. Written in C.

ELSAview ELSAview ACAD Drawing Manager

Hdwe is:	PC	
DBMS is:		
OPSys is:	Win	ELSA America, Inc.
$ Range:	$199	400 Oyster Point Blvd.
Partner:		San Franciso CA 94080-
Mfr:	ELSA AM	Mfg Phone:(
	1-(800)272-ELSA	Distributor:(415)615-7799x

File Control pops up on top of AutoCAD R12

ForReview ForReview

Hdwe is:		
DBMS is:		
OPSys is:		Advance Technology Ctr
$ Range:		22982 Mill Creek Drive
Partner:		Laguna Hills CA 92653-
Mfr:	ATC	Mfg Phone:(
		Distributor:(714)583-9119x

HyperView HyperView Redline, Tech Edit & View

Hdwe is:	WS	Karels ,John ,
DBMS is:		
OPSys is:	SunUnix	TechView Corp.
$ Range:	$695/user	2500 W. Higgins Rd #1271
Partner:		HoffmanEstat IL 60195-
Mfr:	Techview	Mfg Phone:(
		Distributor:(708)490-0066x

Supports many formats

IGESView IGESview for Windows

Hdwe is:	Multi	Hess ,Charlen,A
DBMS is:	N/A	Director of Marketing
OPSys is:	DosUnix	IGES Data Analysis
$ Range:	$795D $1295U	2001 N. Janice Ave
Partner:		Melrose Park IL 60106-
Mfr:	IDA	Mfg Phone:(708)344-1815
		Distributor:(708)344-1815x334

Allows users to manipulate, redline, integrate & validate CAD data. CALS support, CGM Pubs interface, Raster & TIFF support. Markups stored in IGES format.

PreVIEW PreVIEW

Hdwe is:		Vartanian ,Ken ,
DBMS is:		VP Marketing
OPSys is:		Rosetta Technologies
$ Range:		
Partner:		Portland OR
Mfr:	ROSETTA	Mfg Phone:(

Image Viewing & Redlining Package. Working with CMstat

Red-lining Red-lining

Hdwe is:	WS	
DBMS is:		
OPSys is:	Unix	Graphics Systems, Inc.
$ Range:	$10K for 10	
Partner:		
Mfr:	Graphics	Mfg Phone:(410)224-2926

Red-lining is available on HP, IBM, DEC and Sun.
Allows users to view TIFF Group IV images and create CGM vector markup files.

REDLINE Formtek:REDLINE

Hdwe is:	WS	Murray ,Helen ,G
DBMS is:		Mgr, Mktg Communications
OPSys is:	Sun 4+	Formtek, Inc.
$ Range:	$2K+/seat	661 Andersen Dr
Partner:		Pittsburgh PA 15220-9932
Mfr:	FORMTEK	Mfg Phone:(
	1-(800)367-6835	Distributor:(412)937-4946x

Redline works on a PC, requires Windows 3.0

Viewpoint AR-VIEWPOINT

Hdwe is:
DBMS is:
OPSys is:
$ Range:
Partner: Tandem
Mfr: ARRON/ROSS

Natzic ,Walter ,
President & CEO
Arron-Ross Corp.
1132 Indian Springs Dr.
Glendora CA 91740-
Mfg Phone:(818)963-4119
Distributor:(818)963-4119x

Includes Screen Capture and Paint program
Source Code

ADC Aide-De-Camp

Hdwe is: Multi
DBMS is: RDBMS &C
OPSys is: UnixVMS
$ Range:
Partner:
Mfr: SMDS

Cole ,Mary ,
CEO
SW Maint & Dev Sys, Inc.

Concord MA 01742-
Mfg Phone:(
Distributor:(508)369-7398x

X-ADC now had Motif style GUI. (5/93)
Supports DoD 2167A, Mil-Std-486

CASEVision CASEVision/ClearCase Advanced Configuration Mgmt

Hdwe is: Multi
DBMS is: RDBMS
OPSys is: Unix
$ Range:
Partner:
Mfr: Silicon Gr
1-(800)800-7441

Moren ,Wayne ,
Sales Rep
Silicon Graphics
2665 Long Lake Road
Roseville MN 55113-
Mfg Phone:(
Distributor:(612)633-1980x

Have Video

CCC/DM Change & Configuration Control/Development & Maint

Hdwe is: MainF
DBMS is:
OPSys is: UnixVAX
$ Range:
Partner:
Mfr: Softool

Sterne ,Jim ,
Program Manager
Softool Corp
340 South Kellogg Ave
Goleta CA 93117-
Mfg Phone:(805)683-5777
Distributor:(805)683-5777x

Integrated Solutions Methodology (ISM) consulting available. Demo Disk

ClearCase **ClearCase**

Hdwe is:	WS	Zais ,Adam ,
DBMS is:		
OPSys is:	Unix	Atria Software, Inc.
$ Range:	$4000	24 Prime Park Way
Partner:		Natick MA 01760-9594
Mfr:	Atria	Mfg Phone:(
		Distributor:(508)650-5126x

Leans more toware configuration management than source code control
Targeting large scale mission critical projects

CMT **Configuration Management Toolkit**

Hdwe is:		
DBMS is:		
OPSys is:		Expertware, Inc.
$ Range:		3235 Kifer Road
Partner:		Santa Clara CA 95051-0804
Mfr:	EXPERTWARE	Mfg Phone:(408)746-0706
		Distributor:(408)746-0706x

CW/CM **CaseWare/CM 3.0 Software CM system**

Hdwe is:		
DBMS is:	Object	
OPSys is:		CaseWare, Inc.
$ Range:	$4,000+	108 Pacifica
Partner:		Irvine CA 92718-
Mfr:	CaseWare	Mfg Phone:(
		Distributor:(714)453-2200x

CW/PT **CaseWare/PT problem tracking system**

Hdwe is:		
DBMS is:	Object	
OPSys is:		CaseWare, Inc.
$ Range:	$3,500+	108 Pacifica
Partner:		Irvine CA 92718-
Mfr:	CaseWare	Mfg Phone:(
		Distributor:(714)453-2200x

ENDEVOR **Endevor/MVS**

Hdwe is:	Multi	Karczewski,Douglas,F
DBMS is:		Area Sales Manager

OPSys is:	MVS	Legent Corporation
$ Range:		8615 Westwood Center Dr
Partner:		Vienna VA 22182-2218
Mfr:	Legent	Mfg Phone:(
		Distributor:(703)734-9494x

Endevor products manage change in the data center, or in the source code. Automates inventory managment, change control, config mgnt & release mgmt in the IBM MVS

Librarian **Librarian**

Hdwe is:	MainF	
DBMS is:		
OPSys is:	MVS	Computer Assoc Intern'al
$ Range:		711 Stewart Ave
Partner:		Garden City NY 11530-4787
Mfr:	CAI	Mfg Phone:(

Panvalet **Panvalet**

Hdwe is:	MainF	
DBMS is:		
OPSys is:	MVS	Pansophic Systems
$ Range:		
Partner:		
Mfr:	PS	Mfg Phone:(

PCMS **Product Configuration Management System**

Hdwe is:		Murphy ,Steven ,F
DBMS is:	Oracle	VP Sales
OPSys is:		SQL Software, Inc.
$ Range:		8000 Towers Crescent Dr.
Partner:		Vienna VA 22182-
Mfr:	SQL SW	Mfg Phone:(
		Distributor:(703)760-7895x

PVCS **Intersolve - Polytron Version Control System**

Hdwe is:	PC	Ball ,Del ,W
DBMS is:		VP Sales
OPSys is:	DosUnix	Intersolve
$ Range:		1700 NW 167th Place
Partner:		Beaverton OR 97006-
Mfr:	Intersolve	Mfg Phone:(503)645-1150
	1-(800)547-7827	Distributor:(503)645-1150x

Software Engineers development tool, manages versions of files & check in/out w/changes flagging
Windows, NT, Dos, OS/2, AIX, Sun, HP-UX, & SCO UNIX. May be on either FoxBase+ or dBase - Demo Disk

RCS **Revision Control System**

Hdwe is:	WS	
DBMS is:		
OPSys is:	Unix	Berkeley Unix versions
$ Range:		
Partner:		
Mfr:	Berkeley	Mfg Phone:(

SCCS **Source Code Control System**

Hdwe is:	WS	
DBMS is:		
OPSys is:	Unix	AT&T
$ Range:		
Partner:		
Mfr:	AT&T	Mfg Phone:(

Developers can control write access to source files, monitor chgs & record in history file.
SCCS is serial, on 1 person can "checkout" a file at 1 time.

SourceSafe **Source Safe "Version Control System"**

Hdwe is:	PC	Iversen ,Larry ,A
DBMS is:		Partner
OPSys is:	WinDos	One Tree Software
$ Range:		
Partner:		Raleigh NC 27604-
Mfr:	One Tree	Mfg Phone:(
		Distributor:(800)397-2323x

Organizes & manages souce code files, amoung many developers, & tracks changes to files
Runs in a network with DOS, Windows, Windows NT and Macintosh

TeamOne **TeamOne**

Hdwe is:	WS	McGill ,Patrick,
DBMS is:		
OPSys is:	Unix	TeamOne Systems
$ Range:		
Partner:		Santa Clara CA
Mfr:	TeamOneSys	Mfg Phone:(

Targeting large scale mission critical projects

TeamWare **TeamWare**

Hdwe is:	WS	Walster ,William,
DBMS is:		Engineering Manager
OPSys is:	Unix	SunPro
$ Range:	$995	2550 Garcia Avenue
Partner:		MountainView CA 94043-
Mfr:	SunPro	Mfg Phone:(
		Distributor:(408)276-3576x

Developers can track multiple releases & versions graphically. Supports Distributed Computing.
It allows parallel "checkout", detects 2nd engr's chgs & enforces conflict resolution.

Tools

BASIS Dsktop Basis Desktop

Hdwe is:	WS	
DBMS is:	Basis+	
OPSys is:	Unix	Info Dimensions, Inc
$ Range:		
Partner:		
Mfr:	IDI	Mfg Phone:(

Searches both content & document structure. Tracking & safeguarding of documents.

Workflow Sys

FloWare **Plexus Interactive Workflow Software**

Hdwe is:	WS	Dunn ,Gary ,
DBMS is:	Informix	
OPSys is:	Unix	Recognition International
$ Range:		1310 Chesapeake Terrace
Partner:		Sunnyvale CA 94089-
Mfr:	Recognitio	Mfg Phone:(
		Distributor:(408)743-4300x

Graphical mapping of the steps in a process w/serial & parallel routing available.
Uses Scanned/Faxed images. Client/Server based with Windows 3.1 and Mac interfaces. Demo Disk

FlowLogic **FlowLogic**

Hdwe is:		Uzoma ,Adindu ,
DBMS is:	Object	President
OPSys is:	UltrDos	Workflow Systems, Inc.
$ Range:	$10K & up	100 Fifth Ave.
Partner:		Waltham MA 02154-
Mfr:	WrkFlowSys	Mfg Phone:(
		Distributor:(617)487-7945x

A workflow & routing system, DataBase Independant, allows clients to structure their own attributes
New concept, unique implementation. Initial version on DEC Ultrix/ Motif and PC, w/ethernet.

OfficIQ **Office.IQ**

Hdwe is:	PC	
DBMS is:		
OPSys is:	Win	Portfolio TechnologiesInc
$ Range:		
Partner:		Newark CA
Mfr:	Portfolio	Mfg Phone:(

Concurrent info sharing format w/doc's & folders to pass text, spreadsheets, faxes, & scanned
Windows based for NetWare.

Workflow

Hdwe is:		Chalstrom ,Brownel,
DBMS is:		Sr VP Prod Development
OPSys is:		Action Technologies, Inc.
$ Range:		
Partner:		Alameda CA
Mfr:	Action	Mfg Phone:(

Approach is to take the technology and license that to companies to include in all kinds of

Have an OEM agreement to provide to Lotus for use alongside Lotus Notes 3.0

WorkMan **WorkMan**

Hdwe is:		Spies ,Michael,
DBMS is:		VP Marketing
OPSys is:		Reach Software Corp.
$ Range:		
Partner:		Sunnyvale CA
Mfr:	Reach	Mfg Phone:(

Organizes, routes, processes & tracks into over Novell, NetWare or Banyan Sys's VINES

Uses a "store & forward" messaging paradigm.

Manufacturing

CAM

CADAM **CADAM Interactive Design**
Hdwe is: Multi Andrews ,Tony ,
DBMS is:
OPSys is: SecNote CADAM Inc.
$ Range:
Partner: Minneapolis MN
Mfr: CADAM INC Mfg Phone:(
Distributor:(612)854-1584x

Modules of N/C, 3D Interactive, 3D Mesh, Piping, Printed Ckts, Geometry Intfc & Interactive Solids

Runs on IBM MainF, Unix, & PC's. Evolved from Lockheed in-house system in the 60's.

CIM CAM **Numeric Control Programming**
Hdwe is: WS Machut ,Terry ,L
DBMS is: UNIX Sales Mgr, Central Region
OPSys is: Cimlinc, Inc.
$ Range: 1110 N. Olde World St.
Partner: Milwaukee WI 53203-
Mfr: CIMLINC Mfg Phone:(708)250-0090
1-(800)225-7943 Distributor:(414)347-7845x

CIM CUT **Numeric Control Programming**
Hdwe is: WS Machut ,Terry ,L
DBMS is: UNIX Sales Mgr, Central Region
OPSys is: Cimlinc, Inc.
$ Range: 1110 N. Olde World St.
Partner: Milwaukee WI 53203-
Mfr: CIMLINC Mfg Phone:(708)250-0090
1-(800)225-7943 Distributor:(414)347-7845x

CIM SURF **Numeric Control Programming**
Hdwe is: WS Machut ,Terry ,L
DBMS is: UNIX Sales Mgr, Central Region
OPSys is: Cimlinc, Inc.
$ Range: 1110 N. Olde World St.
Partner: Milwaukee WI 53203-
Mfr: CIMLINC Mfg Phone:(708)250-0090
1-(800)225-7943 Distributor:(414)347-7845x

SmartCAM **SmartCAM Integrated CAM Systems**
Hdwe is: Multi
DBMS is:
OPSys is: Unix Great River Sys/MicroAge
$ Range:
Partner:
Mfr: PointCntrl Mfg Phone:(

Advanced 3D machining, prototyping and CNC.
John Parsons, "Father of NC"

GT

Brisch-Birn **Group Technology**
Hdwe is: Hyde ,Bill ,
DBMS is: President
OPSys is: Brisch-Birn & Partners
$ Range: 1656 SE 10th Terrace
Partner: FtLauderdale FL 33316-
Mfr: BBP Mfg Phone:(
Distributor:(305)525-3166x

Classification / Coding system based on numeric coding system
One of the original systems on the market

D-Class **Group Technology**
Hdwe is:
DBMS is:
OPSys is: CAM Software, Inc.
$ Range: 390 West 800 North
Partner: Orem UT 84059-0276
Mfr: CAM Mfg Phone:(801)225-0080
Distributor:(801)225-0080x

HMS-GT **HMS - Group Technology**
Hdwe is: WS Green ,Jim ,
DBMS is: Consultant
OPSys is: Unix Houtzeel Mfg Systems, Inc
$ Range:
Partner: Cimlinc Waltham MA 02254-1605
Mfr: Houtzeel Mfg Phone:(
Distributor:(617)890-2811x

Group Technology analysis system built upon the Cimlinc Linkage Compound Document tool.

OIR **Group Technology**

Hdwe is:
DBMS is:
OPSys is: ITI
$ Range:
Partner: Cinncinati OH
Mfr: ITI Mfg Phone:(

Alex Houtzeel was involved with OIR. OIR to ITI may now be part of Computer Vision

Mfg Doc

HMS-CAPP **HMS - Computer Aided Process Plan**

Hdwe is: WS Green ,Jim ,
DBMS is: Consultant
OPSys is: Unix Houtzeel Mfg Systems, Inc
$ Range: $2,000+
Partner: Cimlinc Waltham MA 02254-1605
Mfr: Houtzeel Mfg Phone:(
Distributor:(617)890-2811x

CAPP process built upon the Cimlinc Linkage Compound Document tool.

Intellicap **Intellicap**

Hdwe is:
DBMS is:
OPSys is: CIMtelligence CAPP
$ Range:
Partner:
Mfr: CIMtellige Mfg Phone:(

Supposedly integrates CAD/CAM - MRPII - & SFC

MODS **Manufacturing Operation Documentation System**

Hdwe is: WS Heller ,Terry ,
DBMS is: Propriet
OPSys is: Apollo DocuGraphix, Inc.
$ Range:
Partner: Chicago IL
Mfr: DOCUGRAPHI Mfg Phone:(
Distributor:(708)351-1891x

Incl: DocuManager (Release); DocuReview; DocuView (Read only); Electronic Shop Documentation (ESD)
Also runs on VAX, HP, PC Xterm.

OASSIS **Operation & Setup Sheets Intelligent System**

Hdwe is:		Peled ,Joseph ,
DBMS is:		Systems Designer
OPSys is:		Gerber Systems Technology
$ Range:		
Partner:		So. Windsor CT
Mfr:	Gerber	Mfg Phone:(

PMW **Paperless Manufacturing Workplace**

Hdwe is:		
DBMS is:		
OPSys is:		IBM
$ Range:		3200 Windy Hill Road
Partner:		Marietta GA 30067-
Mfr:	IBM	Mfg Phone:(
		Distributor:(404)835-8522x

MRPII

4thShift **FOURTH SHIFT**

Hdwe is:	PC	Bush ,Bill ,
DBMS is:		Sales
OPSys is:	DOS	Fourth Shift Corp.
$ Range:	$60K	7900 International Drive
Partner:		Minneapolis MN 55425-
Mfr:	4thSHIFT	Mfg Phone:(
	1-(800)342-5675	Distributor:(612)851-1800x

AMAPS **AMAPS**

Hdwe is:	MainF	Wadden ,Mike ,
DBMS is:	Adabase	Sales
OPSys is:	IBM	Mgmt Sciences America
$ Range:		3400 Yankee Drive
Partner:		Eagan MN 55122--104
Mfr:	MSA	Mfg Phone:(
		Distributor:(612)681-7000x

Avalon

Hdwe is:	Multi	Burell ,Dean ,
DBMS is:	Orac/Syb	Marketing
OPSys is:	Unix	Avalon Software
$ Range:		
Partner:		
Mfr:	Avalon	Mfg Phone:(
		Distributor:(602)570-6849x

BPCS

Hdwe is:		Conley ,Bud ,W
DBMS is:		Sr Apps Consultant
OPSys is:	AS400	System Software Assoc's
$ Range:		325 Hopping Brook Rd
Partner:		Holliston MA 01746-
Mfr:	SSA	Mfg Phone:(
		Distributor:(508)429-7200x

CAS/AD **Cmptr Assoc-CAS/Aerospace & Defense**

Hdwe is:		
DBMS is:	Relat'al	
OPSys is:		Computer Assoc Intern'al
$ Range:		711 Stewart Ave
Partner:		Garden City NY 11530-4787
Mfr:	CAI	Mfg Phone:(
	1-(800)645-3003	

CMS **Cullinet Manufacturing System**

Hdwe is:		LeDuc ,John ,C
DBMS is:	IDMS/R	Account Executive
OPSys is:		Cullinet Software, Inc.
$ Range:		701 4th Ave. So.
Partner:		Minneapolis MN 55415-
Mfr:	CULLINET	Mfg Phone:(617)329-7700
		Distributor:(612)341-2646x

Contr:F **Control: Financial**

Hdwe is:	MainF	Dorece ,Pat ,
DBMS is:	IBM/VAX	
OPSys is:		Cincom Systems, Inc.
$ Range:		2300 Montana Ave.
Partner:		Cincinnati OH 45211-
Mfr:	CINCOM	Mfg Phone:(513)662-2300

1-(800)543-3010

Contr:M **Control: Manufacturing**

Hdwe is: MainF Dorece ,Pat ,

DBMS is: IBM/VAX

OPSys is: Cincom Systems, Inc.

$ Range: 2300 Montana Ave.

Partner: Cincinnati OH 45211-

Mfr: CINCOM Mfg Phone:(513)662-2300

1-(800)543-3010) - x

Has a documentation module

COPICS **COPICS**

Hdwe is: MainF

DBMS is:

OPSys is: IBM Corp.

$ Range:

Partner:

Mfr: IBM Mfg Phone:(

Production Control

ERP **Enterprise Resource Planning**

Hdwe is:

DBMS is: RDBMS

OPSys is: Unix Oracle

$ Range:

Partner:

Mfr: Oracle Mfg Phone:(

ECO module only allows redline and review of the PL.

Expandable **Expandable**

Hdwe is: PC

DBMS is:

OPSys is: Windows Expandable

$ Range:

Partner:

Mfr: Mfg Phone:(

Strong BOM, Inventory, Weak Financials

FlowStream FlowStream
Hdwe is:
DBMS is:
OPSys is: Concilium
$ Range:
Partner: MountainView CA
Mfr: Concilium Mfg Phone:(

Great Plains Great Plains
Hdwe is: MacIn
DBMS is:
OPSys is: Sys 7 Unknown
$ Range:
Partner:
Mfr: Spectrum Mfg Phone:(

Financials only!

GrowthPower Growth Power
Hdwe is:
DBMS is:
OPSys is: Powersoft Corp.
$ Range:
Partner: Burlington MA
Mfr: PowerSoft Mfg Phone:(
Distributor:(617)229-2200x

Input from Kevin Ehringer, Dallas '93

Intreprid Intreprid
Hdwe is:
DBMS is:
OPSys is: NE Data Systems
$ Range:
Partner:
Mfr: NE Data Mfg Phone:(

Change Effectivity by Date or S/N is optional

JIT Resource Just In Time Resources
Hdwe is:
DBMS is: Oracle
OPSys is: International, Inc.
$ Range: 1705 S. Capital of TX Hwy

Partner:		Austin TX 78746-
Mfr:	Internatio	Mfg Phone:(
	1-(800)433-2467	Distributor:(800)433-2467x

Has ECR/ECO & Config Module

Macola

Hdwe is:	PC	
DBMS is:		
OPSys is:	DOS/SCO	Macola, Inc.
$ Range:		333 E. Center Street
Partner:		Marion OH 43302-4148
Mfr:		Mfg Phone:(
	1-(800)468-0834	Distributor:(800)468-0834x550

Runs on PC DOS, OS/2, Windows & SCO/Unix

MacPac **MacPac**

Hdwe is:		
DBMS is:		
OPSys is:		Arthur Anderson
$ Range:		
Partner:		
Mfr:	AA	Mfg Phone:(

Has "Expert Configurator" module

ManBase **ManBase 7**

Hdwe is:		Gould ,Mike ,
DBMS is:	RDBMS	Sales
OPSys is:	UnixDos	MAI Systems Corp.
$ Range:	$60K	104 Decker Ct.
Partner:		Irving TX 75062-
Mfr:	MAI	Mfg Phone:(313)347-9070
		Distributor:(214)717-6757x

Business planning & control system for processor mfgrs, i.e. chemical, food & pharmaceuticals.
Uses Client/Server model. Manage inventory, production, sales orders w/ full financials.

ManMan ManMan

Hdwe is:	Multi	Carlson ,Eric ,
DBMS is:	Ingres	President & CEO
OPSys is:		Ask Group, Inc
$ Range:		2440 W. El Camino Real
Partner:		MountainView CA 94039-7640
Mfr:	ASK	Mfg Phone:(
	1-(800)4FA-CTOR	Distributor:(415)969-4442x

VAX/VMS
OMAR is a separate module

ManManX ManManX (new product)

Hdwe is:	Multi	Carlson ,Eric ,
DBMS is:	Ingres	President & CEO
OPSys is:		Ask Group, Inc
$ Range:	$25K - $1M	2440 W. El Camino Real
Partner:		MountainView CA 94039-7640
Mfr:	ASK	Mfg Phone:(
	1-(800)4FA-CTOR	Distributor:(415)969-4442x

NEW Release, ported to Unix.
A 20 module, 30 user system lists about $200K. Unix, HP, VAX

MAPICS Mfg Actg & Production Information Control Sys

Hdwe is:	MainF	
DBMS is:		
OPSys is:		IBM Corp.
$ Range:		
Partner:		
Mfr:	IBM	Mfg Phone:(

Mfg Control, Order Entry, BOM

Max Max

Hdwe is:	PC	
DBMS is:		
OPSys is:	Windows	Max
$ Range:		
Partner:		
Mfr:		Mfg Phone:(

Will "rent to buy" for 1 year @ $900/mo

MfgPRO MfgPRO

Hdwe is:		Erickson ,Steve ,
DBMS is:	RDBMS	Sales
OPSys is:		QAD Inc.
$ Range:		20415 Howland Ave.
Partner:		Lakeville MN 55044-
Mfr:	QAD	Mfg Phone:(
		Distributor:(612)469-1914x

Progress or Oracle RDBMS

MINXware MINXware

Hdwe is:	WS	Duflo ,Peter ,
DBMS is:	"C"	Sales
OPSys is:	UNIX	Minx Software, Inc.
$ Range:		600 W. Cummings Pkway
Partner:		Woburn MA 01801-
Mfr:	Minx	Mfg Phone:(
		Distributor:(617)932-0932x

Limited ECO implementation tracking by department.

MISYS Mfg Inventory System

Hdwe is:		Merrill ,Terry ,
DBMS is:	"C" Code	
OPSys is:		Micro Computer Specialist
$ Range:	$1495	
Partner:		VT
Mfr:	MCS	Mfg Phone:(
		Distributor:(802)457-4600x

This package works with Computer Associates packages.

Platinum Platinum

Hdwe is:	PC	
DBMS is:		
OPSys is:	Windows	Platinum
$ Range:		
Partner:		
Mfr:		Mfg Phone:(

Weak BOM, Inventory Strong Financials

PPS/AWI **Process Planning & Auto Work Instruction System**

Hdwe is:
DBMS is:
OPSys is:
$ Range:
Partner: Tandem
Mfr: ARRON/ROSS

Natzic ,Walter ,
President & CEO
Arron-Ross Corp.
1132 Indian Springs Dr.
Glendora CA 91740-
Mfg Phone:(818)963-4119
Distributor:(818)963-4119x

Symix **Symix**

Hdwe is: Multi
DBMS is: Progress
OPSys is: UnixDos
$ Range:
Partner:
Mfr: Symix

Eskin ,Susan ,E
VP of Marketing
Symix Cmptr Systems, Inc.
2800 Corporate Exchange D
Columbus OH 43231-
Mfg Phone:(
Distributor:(614)523-7000x

Includes 15 Mfg and Financial modules. Has ACAD BOM xfer & Mfg ECN Module

Supports VMS, Unix & DOS. Runs on 400 platforms. Progress is a RDBMS

TelesisMFG **Telesis**

Hdwe is:
DBMS is: RDBMS
OPSys is: Unix
$ Range:
Partner:
Mfr: Telesis

Mueller ,David ,E
Consulting Mgr
Telesis Computer Corp.
207 Sigma Drive
Pittsburg PA 15238-
Mfg Phone:(
Distributor:(412)963-8844x

UnisD **Unisys Defense MRPII System**

Hdwe is:
DBMS is:
OPSys is:
$ Range:
Partner:
Mfr: Cimcase

ACME Engineering
Mfg Phone:(

Cimcase is a division of ACME Engineering
D stands for Defense applications

WDS **COMPASS CONTRACT**

Hdwe is:
DBMS is:
OPSys is: Western Data Systems
$ Range:
Partner:
Mfr: WDS Mfg Phone:(

Combination CM / MRP system, does not cover all of CM.
I could use more info, a contact or data sheets on this.

Other

DB - 4GL

CS/ADS **Convergent Solutions/Application Development Sys**

Hdwe is: Multi
DBMS is: SQL
OPSys is: DosUnix — Convergent Solutions, Inc
$ Range:
Partner:
Mfr: Convergent — Mfg Phone:(

4GL works with Oracle, InterBase, Informix, Sharebase; Openlook, Motif or Windows
Has Data Dictionary, Report Generator, Forms Painter, Graphs, & Procedural Def Language

DB Designer **Cadre DB Designer**

Hdwe is: Multi — Chapman ,Robin ,
DBMS is: RDBMS — Supv, Telemarketing
OPSys is: — Cadre Technologies Inc.
$ Range: — 222 Richmond Street
Partner: — Providence RI 02903-
Mfr: Cadre — Mfg Phone:(
Distributor:(401)351-5950x

Focus **Focus**

Hdwe is:
DBMS is:
OPSys is: SCOUnix — Information Builders, Inc
$ Range: $995
Partner:
Mfr: InfoBldrs — Mfg Phone:(

Single View Reporting & Common 4GL across Informix, Oracle, Sybase, C-ISAM, ODT Data, Ingres
The foundation for an Open Applications Development environment. Focus applic's are portable.

UNIFACE **UNIFACE 4GL**

Hdwe is: Multi
DBMS is: SQL
OPSys is: Unix — Uniface B.V.
$ Range:

Partner:
Mfr: Uniface Mfg Phone:(
1-(800)365-3608

4GL works with Oracle, Informix, Sybase, Ingres, Rdb, RMS & others; Openlook, Motif, Windows &
Has 3-schema architecture, FastForm w/multitables

DB - Tools

IQ **IQ Intelligent Query**

Hdwe is:	Multi	Chitty ,Rick ,
DBMS is:	SQL	President
OPSys is:		IQ Software Corp.
$ Range:		3295 River Exchange Drive
Partner:		Norcross GA 30092-9909
Mfr:	IQ	Mfg Phone:(
	1-(800)458-0386	Distributor:(404)446-8880x

Point & click Report Writer, supports UNIX, DOS, VAX/VMS & Windows 3.x

ObjVision **ObjectVision 2.0**

Hdwe is:	PC	Fauntleroy,Bill ,
DBMS is:		Midwest Sales Rep
OPSys is:	Win3.x	Borland International Inc
$ Range:	$149	
Partner:		Chicago IL
Mfr:	Borland	Mfg Phone:(
	1-(800)331-0877	Distributor:(708)678-2288x

Access Paradox, dBASE, Btrieve & ACSII. Draft-Pac BOM generator???
Supports Dynamic Link Libs, Dynamic Data Exchange, & Object Linking & Embedding.

PersAccess **Personal Access**

Hdwe is:	PC	
DBMS is:		
OPSys is:	Win3.0	Spinnaker Software Corp.
$ Range:		
Partner:		
Mfr:	Spinnaker	Mfg Phone:(

Database Query/Report tool. Lets users without SQL access corporate databases
Supports Oracle, SQL Server, Sybase, Paradox, dBASE, & Novells Netware Btrieve

Netware Btrieve. w/multimedia

PowerBuilder PowerBuilder (also PowerMaker)

Hdwe is:	PC	
DBMS is:	WatcomRD	
OPSys is:	Win 3.1	Powersoft Corp.
$ Range:	$349	
Partner:		Burlington MA
Mfr:	PowerSoft	Mfg Phone:(
		Distributor:(617)229-2200x

A complete, object-oriented form dev environ; repository & db mgmt, query, report & graphs.
Tool to build forms without programming, not a CM application. [PowerBuilder w/OOP is $3,995]

PowerMaker PowerMaker (also PowerBuilder)

Hdwe is:	PC	
DBMS is:	WatcomRD	
OPSys is:	Win 3.1	Powersoft Corp.
$ Range:	$349	
Partner:		Burlington MA
Mfr:	PowerSoft	Mfg Phone:(
		Distributor:(617)229-2200x

A complete, object-oriented form dev environ; repository & db mgmt, query, report & graphs.
Tool to build forms without programming, not a CM application. [PowerBuilder w/OOP is $3,995]

SQLWin SQLWindows

Hdwe is:		
DBMS is:	SQL	
OPSys is:	DOS	Gupta Technologies, Inc.
$ Range:	$1995	
Partner:		Menlo Park CA
Mfr:	Gupta	Mfg Phone:(

Multi platform DOS SQL database

uniface uniface cross-platform development environment

Hdwe is:	Multi	Shukla ,Anu ,
DBMS is:	Multi	VP Worldwide Marketing
OPSys is:	Multi	Uniface Corp.
$ Range:		1320 Harbor Bay Parkway
Partner:		Alameda CA 94501-

Mfr: Uniface Mfg Phone:(
Distributor:(800)365-3608x

Model-driven dev environment seamlessly integrates data from multiple RDBMS's across many
Supports Oracle, ALLBASE/SQL, Sybase, Ingres, Informix, TurboImage on Windows, Motif, OS/2,

VisualBasic **Visual Basic**
Hdwe is: PC
DBMS is:
OPSys is: Win 3.1 Microsoft
$ Range:
Partner:
Mfr: MS Mfg Phone:(
1-(800)458-0386

Windows application builder, uses the Windows Graphical User Interface. Is a tool to build applications, not a CM application. Can interface to SQL DBMS's, OLE & Images.

DBMS's

ACCESS **ACCESS**
Hdwe is: PC
DBMS is: RDBMS
OPSys is: Windows Microsoft
$ Range: $495/$99
Partner:
Mfr: Microsoft Mfg Phone:(

BASISplus **Basis Plus Database Manager**
Hdwe is: WS
DBMS is: Basis+
OPSys is: Unix Info Dimensions, Inc
$ Range:
Partner:
Mfr: IDI Mfg Phone:(

DB2

Hdwe is:	MainF	
DBMS is:	SQL	
OPSys is:		IBM
$ Range:		3200 Windy Hill Road
Partner:		Marietta GA 30067-
Mfr:	IBM	Mfg Phone:(
		Distributor:(404)835-8522x

IBM mainframe RDBMS

dBaseIV **dBase IV (Ashton-Tate)**

Hdwe is:	Multi	Fauntleroy,Bill ,
DBMS is:	SQL	Midwest Sales Rep
OPSys is:	UnixDos	Borland International Inc
$ Range:		
Partner:		Chicago IL
Mfr:	Borland	Mfg Phone:(
		Distributor:(708)678-2288x

dBase for Windows "field test" version
Rpt Card 6.06, SW Security low, Data Manipulation high.

EMPRESS **Empress**

Hdwe is:		Kornatowsk,John ,
DBMS is:	SQL	President
OPSys is:		Empress Software
$ Range:		
Partner:		
Mfr:	Empress	Mfg Phone:(301)220-1919
		Distributor:(301)220-1919x

FoxPro **FoxPro & FoxPro for Windows**

Hdwe is:	PC	
DBMS is:	RDBMS	
OPSys is:	WinDos	Microsoft
$ Range:		
Partner:		
Mfr:	Microsoft	Mfg Phone:(

INFORMIX **Informix Relational Database Management System**

Hdwe is:	Multi	Tolvstad ,Eric ,
DBMS is:	SQL	Account Coordinator
OPSys is:	Multi	Great River Sys

$ Range:		4252 Park Ave
Partner:		Mpls MN 55407-
Mfr:	Informix	Mfg Phone:(415)926-6300
		Distributor:(612)686-0995x 10

Highly rated RDBMS
Rpt Card 6.66, Rpt Writer low, Ease of data retreival high.

INGRES **Ingres Relational Database Managment System**

Hdwe is:	Multi	Casey ,Mark ,L
DBMS is:	SQL	Account Marketing Rep
OPSys is:	Multi	Ingres, An ASK Company
$ Range:		7760 France Ave. So.
Partner:		Minneapolis MN 55435-
Mfr:	Ingres	Mfg Phone:(510)769-1400
	1-(800)4IN-GRES	Distributor:(612)831-0664x

Good, technically solid DBMS
Current Ver 6.3 in 1991. Rpt Card 5.38, Rpt Writer low, Ease of Use high.

Interbase **Interbase Software**

Hdwe is:	WS	Fauntleroy,Bill ,
DBMS is:	RDBMS	Midwest Sales Rep
OPSys is:	Unix	Borland International Inc
$ Range:		
Partner:		Chicago IL
Mfr:	Borland	Mfg Phone:(
		Distributor:(708)678-2288x

DB Server, interfaces with Object Vision, Paradox, Quattro Pro,
SQL, concurrent db w/Referential Integrity & business rules in datadictionary. Started in '86

Montage **Montage Object-Relational Database Management Sys**

Hdwe is:	WS	Golden ,Bruce ,
DBMS is:	O-RDBMS	Director of Marketing
OPSys is:	Multi	Montage Software
$ Range:	$995 /user	
Partner:		Emeryville CA
Mfr:	Montage	Mfg Phone:(
		Distributor:(510)652-8000x

a hybrid Object-Relational database allows SQL queries of graphical objects
Able to manage heterogeneous collections of information, traditional data to images, sound & video.

ObjStore **ObjectStore**
Hdwe is:
DBMS is: OODBMS
OPSys is: Object Design
$ Range:
Partner: Burlington MA
Mfr: ObjectDes Mfg Phone:(

OODBMS
Hdwe is: Miezwa ,Sandy ,
DBMS is: OODBMS Sales Representive
OPSys is: Itasca Systems, Inc.
$ Range: 7850 Metro Parkway
Partner: Minneapolis MN 55425-
Mfr: Itaska Mfg Phone:(
Distributor:(612)851-3169x

ORACLE **Oracle 7.1**
Hdwe is: Multi Becker ,Bob ,E
DBMS is: SQL District Sales Manager
OPSys is: Multi Oracle Corp.
$ Range: 8500 Normandale Lake Blvd
Partner: Minneapolis MN 55437-
Mfr: Oracle Mfg Phone:(
Distributor:(612)835-9266x
Rpt Card 6.35, Memory Reqmts low, Sorting capability high.

PARADOX **Paradox**
Hdwe is: PC Fauntleroy,Bill ,
DBMS is: RDBMS Midwest Sales Rep
OPSys is: WinDos Borland International Inc
$ Range:
Partner: Chicago IL
Mfr: Borland Mfg Phone:(
Distributor:(708)678-2288x

PROGRES **Progress, Relational DBMS**
Hdwe is: Crismond ,Michael,J
DBMS is: SQL VP No. American Sales
OPSys is: Progress Software Corp.
$ Range: 14 Oak Park
Partner: Bedford MA 01730-USA
Mfr: Progress Mfg Phone:(

Distributor:(617)280-4000x

Rpt Card 8.34, Memory Reqmts low, data manipulation high. #1 in all features

R:BASE **R:Base Relational DBMS Rel 4.5**

Hdwe is:	PC	R:BASE ,1 ,
DBMS is:		
OPSys is:	DOS	Microrim, Inc.
$ Range:	$795	3925 159th NE
Partner:		Redmond WA 98052-
Mfr:	Microrim	Mfg Phone:(
	1-(800)248-2001	Distributor:(800)762-5240x40E

probably the most complete relational model for DOS.
Indexes & validation rules are built into the database structure.

SQL Server **SQL Server**

Hdwe is:		
DBMS is:	Sybase	
OPSys is:		Microsoft Corp.
$ Range:		
Partner:		
Mfr:	Microsoft	Mfg Phone:(

SQLBase **SQL Database Server**

Hdwe is:	Multi	Gupta ,1 ,
DBMS is:	SQL	
OPSys is:	UnixDos	Gupta Technologies, Inc.
$ Range:		
Partner:		Menlo Park CA
Mfr:	Gupta	Mfg Phone:(

SUPRA **SUPRA RDBMS**

Hdwe is:		Dorece ,Pat ,
DBMS is:	SQL	
OPSys is:	Multi	Cincom Systems, Inc.
$ Range:		2300 Montana Ave.
Partner:		Cincinnati OH 45211-
Mfr:	Cincom	Mfg Phone:(

Client/Server.

SYBASE **Sybase**

Hdwe is:		EpsteinPhD,Robert ,
DBMS is:	SQL	Exec Vice President

OPSys is: Unix — Sybase, Inc.
$ Range: — 6475 Christie Ave.
Partner: — Emeryville CA 94608-
Mfr: Sybase — Mfg Phone:(
1-(800)8SY-BASE — Distributor:(800)8SY-BASEx

Program Mgmt

AutoPlan II — **AutoPlan II Project Management Tools**

Hdwe is: WS — Whaten Jr.,Phillip,F
DBMS is: — VP Sales/Marketing
OPSys is: Unix — Digital Tools
$ Range: $1495 & up — 18900 Stevens Creek Blvd
Partner: — Cupertino CA 95014-
Mfr: Digital — Mfg Phone:(
1-(800)755-0065 — Distributor:(408)366-6920x

Provides Ghantt & Pert Charts, Like a Timeline or MS Project
Runs on Sun Sparc and HP workstations

AWC — **Auto Write & Send Correspondence**

Hdwe is: — Natzic ,Walter ,
DBMS is: — President & CEO
OPSys is: — Arron-Ross Corp.
$ Range: — 1132 Indian Springs Dr.
Partner: Tandem — Glendora CA 91740-
Mfr: ARRON/ROSS — Mfg Phone:(818)963-4119
Distributor:(818)963-4119x

ProjMgr — **Project Manager**

Hdwe is: PC
DBMS is:
OPSys is: DOS/WIN — Microsoft
$ Range:
Partner:
Mfr: Microsoft — Mfg Phone:(

TimeLine — **TimeLine Project Manager**

Hdwe is: PC — Taylor ,Ellen ,
DBMS is: — VP and GM
OPSys is: DOS/WIN — Symantec
$ Range:
Partner: — Santa Monica CA

Mfr: Symantec Mfg Phone:(

TimeSlips **Timeslips**
Hdwe is: PC
DBMS is:
OPSys is: DOS/WIN Timeslips
$ Range:
Partner:
Mfr: Mfg Phone:(

Publications

AVPS **Avalon Publisher**
Hdwe is: Lee ,Sheila ,
DBMS is: Account Manager
OPSys is: Unix Qualix Group, Inc.
$ Range: $995 1900 S. Norfolk St
Partner: San Mateo CA 94403-1151
Mfr: Avalon Mfg Phone:(
Distributor:(415)572-0200x

CONTEXT
Hdwe is: WS
DBMS is:
OPSys is: Mentor Context
$ Range:
Partner:
Mfr: CONTEXT Mfg Phone:(

Word Processor, maintains change history (adds/deletes) in the doc file. Can recreate a specific

DocEXPRESS DocEXPRESS 2.0 Intelligent Auto-documentation

Hdwe is:	WS	Nichols ,Yukiko ,
DBMS is:		Sales Support
OPSys is:	Unix	ATA Inc.
$ Range:		3528 Torrance Blvd
Partner:		Torrance CA 90503-
Mfr:	ATA	Mfg Phone:(
		Distributor:(310)316-6350x

Assists system & SW engr's develop documents per reqmt's of DoD & others.
Is integrated with Unix CASE tools & Document Publishing Systems, FrameMaker & Interleaf

DynaText DynaText Electronic Book Publishing System

Hdwe is:	Multi	
DBMS is:		
OPSys is:	X Windo	Electronic Book Technolog
$ Range:	$1 - $10K	
Partner:		
Mfr:	EBT	Mfg Phone:(401)421-9550

Automatically creates electronic books using any SGML source document. Includes redlining.
Supports SGML but doesn't convert.

FrameBuilder FrameBuilder

Hdwe is:	WS	Schmidt ,Laurie ,
DBMS is:		Sales Representative
OPSys is:	UNIX	Frame Technology Corp.
$ Range:	$2500	1010 Rincon Circle
Partner:		San Jose CA 95131-
Mfr:	FrameTech	Mfg Phone:(
	1-(800)U4F-RAME	Distributor:(408)922-2725x

WYSIWYG Structured Document builder. Supports SGML
The Compound Doc Gateway, path to import text, graphics, audio & video files.

FrameMaker FrameMaker Electronic Publishing

Hdwe is: WS Schmidt ,Laurie ,
DBMS is: Sales Representative
OPSys is: UNIX Frame Technology Corp.
$ Range: 1010 Rincon Circle
Partner: San Jose CA 95131-
Mfr: FrameTech Mfg Phone:(
1-(800)U4F-RAME Distributor:(408)922-2725x
WYSIWYG Electronic Publishing

FrameViewer FrameViewer Electronic Document Distribution SW

Hdwe is: WS Schmidt ,Laurie ,
DBMS is: Sales Representative
OPSys is: UNIX Frame Technology Corp.
$ Range: 1010 Rincon Circle
Partner: San Jose CA 95131-
Mfr: FrameTech Mfg Phone:(
1-(800)U4F-RAME Distributor:(408)922-2725x
Hypertext Navigation, Multimedia, view only (w/formatting), graphics supported, search & retrieval
Has "post-it" notes, electronic bookmarks, API interface.

Illus Illustrator

Hdwe is: PC
DBMS is:
OPSys is: DOS-MAC Adobe Systems
$ Range:
Partner:
Mfr: Adobe Mfg Phone:(

IntelliTAG IntelliTAG SGML

Hdwe is: WS
DBMS is:
OPSys is: Unix Word-Perfect Corp.
$ Range: $495 / user 1555 N. Technology Way
Partner: Orem UT 84057-
Mfr: WordPerf Mfg Phone:(
Distributor:(801)225-5000x
Converts word processing docs to SGML.

INTERLEAF Interleaf 6 Publisher

Hdwe is:	WS	Hanzlik ,Vern ,
DBMS is:		Sales
OPSys is:		Tech Pub Solutions, Inc.
$ Range:	$2,500 per user	5500 Lincoln Drive
Partner:		Minneapolis MN 55436-
Mfr:	Interleaf	Mfg Phone:(
		Distributor:(612)938-4490x

Supports SGML, spell checker, text rotation, links to SQL, interactive equations, multipage tables.
Supports workgroups, users can share doc's and track revisions.

PageMaker PageMaker

Hdwe is:	PC	
DBMS is:		
OPSys is:	DOS-MAC	Adlus Corp.
$ Range:	$519	
Partner:		
Mfr:	Aldus	Mfg Phone:(

Popular page layout program for Macintosh and IBM compatibles.

Ventura Corel Ventura Publisher

Hdwe is:	PC	
DBMS is:		
OPSys is:	DOS-MAC	
$ Range:		
Partner:		
Mfr:	Corel Corp.	Mfg Phone:(

WorldView WorldView V2.0, Interleaf Document Viewer

Hdwe is:	WS	Hanzlik ,Vern ,
DBMS is:	LISP	Sales
OPSys is:	Unix	Tech Pub Solutions, Inc.
$ Range:	Est$300/seat	5500 Lincoln Drive
Partner:		Minneapolis MN 55436-
Mfr:	Interleaf	Mfg Phone:(
	1-(800)223-6638	Distributor:(612)938-4490x

View native Interleaf files & files that have been converted via WorldPress

to Interleaf format
Has SGML, Outline navigator, Redline.

TQM

QMS **Quality Management System**

Hdwe is:		Natzic ,Walter ,
DBMS is:		President & CEO
OPSys is:		Arron-Ross Corp.
$ Range:		1132 Indian Springs Dr.
Partner:	Tandem	Glendora CA 91740-
Mfr:	ARRON/ROSS	Mfg Phone:(818)963-4119
		Distributor:(818)963-4119x

SPC **Applied Stats**

Hdwe is:	Apple	James ,Bill ,
DBMS is:		
OPSys is:		Applied Statistics Inc.
$ Range:		3080 Centerville Road
Partner:		Little Canad MN
Mfr:		Mfg Phone:(612)481-0202
		Distributor:(612)481-0202x

SPC **Statistical Process Control**

Hdwe is:		Natzic ,Walter ,
DBMS is:		President & CEO
OPSys is:		Arron-Ross Corp.
$ Range:		1132 Indian Springs Dr.
Partner:	Tandem	Glendora CA 91740-
Mfr:	ARRON/ROSS	Mfg Phone:(818)963-4119
		Distributor:(818)963-4119x

Productivity for Business

File Control

PCDocsOpen PCDocs Open. Document Organization & Control Sys.

Hdwe is:	Multi	AllenArtal,Marcia ,
DBMS is:	SQLservr	Mgr, Sales Operations
OPSys is:	Win 3.x	PC DOCS, Inc.
$ Range:		124 Marriott Drive
Partner:		Tallahassee FL 32301-
Mfr:	PC DOCS	Mfg Phone:(
		Distributor:(904)942-3627x

Provides database oriented PC file Document Management. Helps organize & locate files.
C++ Object Oriented. Database Independant. W/Full Text search.

Image Sys

CMI

Hdwe is:		CMI ,1 ,
DBMS is:		
OPSys is:	DOS	CMI
$ Range:		6024 Blue Circle Drive
Partner:		Mtka MN 55343-
Mfr:	CMI	Mfg Phone:(
		Distributor:(612)939-9077x

Optical based image and viewing systems. VERY fast response. Pan/ Zoom etc.

E-Quip Alacrity Desktop Document Manager

Hdwe is:	PC	Folts ,Jim ,R
DBMS is:		President
OPSys is:	Win 3.1	Alacrity Systems Inc.
$ Range:	$1,995	43 Newburg Road
Partner:		Hackettstown NJ 07840-USA
Mfr:	Alacrity	Mfg Phone:(
		Distributor:(908)813-2400x

Image filing system, fax & copier incorporated in you PC or network server.
With a scanner you can store, find, copy, e-mail or fax any paper item on your desk.

FileNet **FileNet document-image processing systems**

Hdwe is:	Multi	Libit ,Jordan ,
DBMS is:		VP Marketing
OPSys is:	DosUnix	FileNet Corp.
$ Range:	$1K to $250K	
Partner:		Costa Mesa CA
Mfr:	Filenet	Mfg Phone:(714)966-3400
		Distributor:(714)966-3400x

Sys's are based on client/server arch. WorkShop 3.0 creates forms & scripts for business tasks

WorkForce Desktop 3.0, Windows, query & view images on PC.

OSAR **FileNet Optical Storage & Retreival libraries**

Hdwe is:	Multi	Libit ,Jordan ,
DBMS is:		VP Marketing
OPSys is:	DosUnix	FileNet Corp.
$ Range:	$1K to $250K	
Partner:		Costa Mesa CA
Mfr:	FileNet	Mfg Phone:(714)966-3400
		Distributor:(714)966-3400x

Sys's are based on client/server arch. WorkShop 3.0 creates forms & scripts for business tasks

WorkForce Desktop 3.0, Windows, query & view images on PC.

TMC/IMAGE **TMC/IMAGE**

Hdwe is:	WS	,.. ,
DBMS is:	a RDBMS	
OPSys is:	Unix	Information Solutions
$ Range:	$50K - $1M	
Partner:		Englewood CO
Mfr:	InfoSolut	Mfg Phone:(303)694-9180
		Distributor:(303)694-9180x

Archives & retrieves documents, bills, forms (handwritten) & organizes, catalogues & stores data.

A Unix-based doc-imaging SW sys, works with networked PCs & X terminals

Workflow Sys

Notes **Lotus Notes**

Hdwe is:		Spindel ,Joan ,
DBMS is:		Group Marketing Mgr
OPSys is:	DOS	Lotus Development Corp.
$ Range:		55 Cambridge Pkwy.
Partner:		Cambridge MA 02142-
Mfr:	Lotus	Mfg Phone:(

onGO **onGO office automation suite.**

Hdwe is:	Multi	Demers ,Mark ,
DBMS is:		Dir Channel Marketing
OPSys is:		Uniplex Integration Sys.
$ Range:		600 E. Las Colinas Blvd.
Partner:		Irving TX 75039-
Mfr:	Uniplex	Mfg Phone:(

Includes "Document Management System (DMS)"

WorkFlo **FileNet WorkFlo Business System**

Hdwe is:	Multi	Libit ,Jordan ,
DBMS is:		VP Marketing
OPSys is:	DosUnix	FileNet Corp.
$ Range:	$1K to $250K	
Partner:		Costa Mesa CA
Mfr:	FileNet	Mfg Phone:(714)966-3400
		Distributor:(714)966-3400x

Sys's are based on client/server arch. WorkShop 3.0 creates forms & scripts for business tasks

WorkForce Desktop 3.0, Windows, query & view images on PC.

Index

9 780824 795740